D0930240

Algorithmics for VLSI

International Lecture Series in Computer Science

These volumes are based on the lectures given during a series of specially funded chairs. The International Chair in Computer Science was created by IBM Belgium in co-operation with the Belgium National Foundation for Scientific Research. The holders of each chair cover a subject area considered to be of particular relevance to current developments in computer science.

The Correctness Problem in Computer Science (1981)
R.S. BOYER and J. STROTHER MOORE

Computer-aided Modelling and Simulation (1982)
J.A. SPRIET and G.C. VANSTEENKISTE

Probability Theory and Computer Science (1983)
G. LOUCHARD and G. LATOUCHE

New Computer Architectures (1984)
J. TIBERGHIEN

Algorithmics for VLSI (1986)
C. TRULLEMANS

Algorithmics for VLSI

Edited by

C. Trullemans
Laboratoire de Microelectronique
Université Catholique de Louvain
Louvain-La-Neuve
Belgium

1986

ACADEMIC PRESS
Harcourt Brace Jovanovich, Publishers
London Orlando San Diego New York Austin
Boston Toronto Sydney Tokyo

ACADEMIC PRESS INC. (LONDON) LTD.
24–28 Oval Road
London NW1 7DX

U.S. Edition published by
ACADEMIC PRESS INC.
Orlando, Florida 32887

British Library Cataloguing in Publication Data

Algorithmics for VLSI

ISBN 0-12-701230-3

Photoset by Paston Press, Norwich and printed by
St Edmundsbury Press Ltd., Bury St Edmunds, Suffolk

Contributors

F. Anceau BULL Systèmes, Les Clayes St Bois, F-78340 France
R. M. Lea Brunel University, Uxbridge, Middlesex UB8 3PH, England
Th. Lengauer FB 10, Universität des Saarlandes, D-6600 Saarbrücken, Federal Republic of Germany
K. Mehlhorn FB 10, Universität des Saarlandes, D-6600 Saarbrücken, Federal Republic of Germany

Preface

The first integrated circuit was developed in July 1958 by J. Kilby. This was an equally important step in the history of both computer science and electronics. It was also the starting point of an evolution towards a field of common interest for computer scientists and electronic engineers: algorithmics for very large scale integrated circuits.

Bridging the gap between algorithmics and circuits is possible thanks to the level of maturity reached today by integrated circuit technology. The introduction of FORTRAN compilers, operating systems, and transistor logic circuits used in computers began in 1958.

As compared to vacuum tubes devices, transistor circuits exhibit a drastic reduction in size and power requirement, and a corresponding increase in reliability. The major problem in attempting a further size reduction is to develop a better interconnection technology. This is basically the aim of integrated circuit technology. This problem was of prime importance in the late 1950s and early integrated MOS logic networks were patented in 1957.

Similarly, the need for a telephone exchange system without any moving mechanical part was the driving force behind the quest for the bipolar transistor, discovered by Schockley, Brattain and Bardeen, who were awarded the Nobel prize in 1949. The discovery of the bipolar transistor is an example of successful interaction between formerly separated fields. Schockley, Brattain and Bardeen had been taught solid state physics by Schrödinger, Franck, Sommerfeld and Schottcky, who were among the founders of this new science; through them, a direct link was established between the fundamental research of the physicists in the 1930s and the integrated circuit by Kilby at the end of the 1950s.

The well-known TTL series is another step in the same direction. It is a set of electronic components. However, the behaviour of TTL components is mostly described using Boolean equations: they are a materialization of abstract mathematical entities, ideal logic operators. This way of looking at them is far removed from solid state physics. However, a TTL circuit is built in such a way as to allow this abstract description of its behaviour, without any care for the electrical details. A TTL circuit is, nevertheless, fairly simple. Progress in fabrication technology has allowed for much more powerful primitives to be realized. A chip currently contains several tens of thousands of transistors.

To organize such a complex system, or to write software on such a large scale, are similar problems. Coming from simple circuit to complex systems, electronic engineers had to join computer scientists in translating onto chips the architecture of minicomputers.

The celebrated – and debated – book, *Introduction to VLSI Systems* by C. Mead and L. Conway, is a sign of a corresponding interest from the computer scientist community. This book is merely an indicator of the emergence of this common interest in VLSI.

The catalog of TTL components was the marker of a previous meeting point. A large set of machines have been built from this restricted set of primitives. The possibilities opened by the VLSI are, however, much more fascinating.

An algorithm handles a set of data in order to produce a solution. Choosing the better method is merely a trade-off between the cost and the performance of the solution. The context of VLSI is, however, very different from the context of classical machines, and this remark opens the way for a whole class of new algorithms.

The new possibilities offered to the computer by the VLSI will obviously change the way computers are constructed, and they will also change what computers are doing. There is obviously a large qualitative difference – not only a quantitative one – between the computation of mathematical tables, which was the task of the first computers, and the job of an office automation system.

Integrated circuit technology is naturally oriented towards large volume producton. It may be compared to printing works. An observer of the 15th century, even if able to forecast the unbelievable quantity of printed paper which is distributed today, would probably not have foreseen the consequence of this flow of texts on a now well-read mankind. Our own ability to forecast is no better. We can only guess that an increasing quantity of algorithmic machines will spread everywhere. At least one may conjecture that the undue standardization issued by the industrial revolution can be cured thanks to information technology.

At the present time, however, standardization is even an internal law of VLSI technology itself: general purpose microprocessors and memories share the larger part of the market. However, custom chips are gaining in importance; they could reach 90% of the market at the end of the 1980s. This is a fundamental trend, explained by the maturation of design techniques, and by the evolution of the processing equipment. A process line is more and more crowded by robots, able to modify their behaviour when facing new problems. A classical bonding machine, driven by cams, can only handle standard chips. Any modification to the bonding-pattern asks for costly mechanical adjustments. An intelligent robot, driven by a pattern recognition system, can handle custom chips at no extra cost.

Moreover, chips themselves are just part of the story: the largest producer of integrated circuits in the world is IBM; however, IBM does not sell integrated circuits, but computers. A VLSI is merely a component of a larger system; the full power of VLSI is to make feasible complex systems at low cost. Even if internally complex, these systems may externally look quite simple. In this way, many problems have a VLSI solution, making full use of the intrinsic capabilities of VLSI technology.

Part 1 of this book (Prof. Lea) introduces the basic concepts of VLSI architectures. Original implementations of special purpose architectures are then described, including a performance comparison between several design styles (e.g. single instructions single data – SISD, and single instruction multiple data – SIMD). In part 2 (Profs. Lengauer and Mehlhorn), a theoretical model for complexity for VLSI is developed and applied to the design of efficient VLSI algorithms. The HILL design system, including a symbolic layout editor and a switch level simulator are also described. Part 3 (Prof. Anceau) is devoted to a detailed analysis of layout design styles. This analysis leads to the description of processor templates, well suited for silicon compilation (part 4). Part 5 summarizes the evolution of processor architecture up to the VLSI era.

The design of complex VLSI systems is adequately performed by borrowing techniques commonly used in computer science, and conversely the versatility of the VLSI is such that it allows for an efficient implementation of complex functions. The algorithmics for VLSI is a promising domain of research, and is likely to be added as a new chapter to the core of computer science. The contribution of the authors of this book is therefore especially welcomed in this series which is devoted to subject areas of particular relevance to computer science.

This book covers the material of a series of lectures given at Louvain-la-Neuve during the academic year 1982–83. The contributors to this book were the holders of the international professorship in computer sciences, organized by the "Fonds de la Recherche Scientifique", which is promoting high-level research and eduction in Belgium. IBM Belgium is generously funding this professorship. I am pleased to express my gratitude to these organizations, to which the attendees to the lectures and the readers of this book are indebted also. I feel also sincerely thankful to the people who helped in the organization of this series of lectures, and especially to Mrs M. Mercenier and Mr M. Windael.

CH. TRULLEMANS

Contents

1. VLSI parallel-processing chip architecture

R.M. LEA

1 INTRODUCTION TO CHIP ARCHITECTURE

In less than 25 years the concept of the integrated circuit has matured from 1 transistor on a single silicon chip to approach 1 000 000 transistors per chip; progressing through SSI (up to 100), via MSI (100 to 1000) and LSI (1000 to 100 000) to VLSI (greater than 100 000). A major contribution to this explosive growth has been a remarkable "collaboration" between the microelectronics component industry and the computer manufacturing industry.

The progress of circuit integration can be likened to that of a goods train with a locomotive at each end, representing "technology push" and "systems pull"; the former demonstrating the eagerness of the semiconductor industry to improve its art and the latter demonstrating the reality of market forces. Clearly, rapid progress can be made only when both "locomotives" work in harmony.

Despite occasional periods of slow progress, this dual propulsion system has rushed the "train" along the "rails" connecting SSI, MSI and LSI into the unexplored region of VLSI. At this point (circa 1980), it became clear that VLSI was not just "a little further down the track" and that the coordination of "technology push" (which had made VLSI possible) and "systems pull" (towards smaller, faster and cheaper computers) had become a major challenge.

The availability of SSI packages incorporating a few logic gates (e.g. quad 2-input NAND gates) enabled digital systems engineers to implement logical functions on a single printed ciruit board, without understanding the underlying semiconductor technology. Thus the traditional "bottom-up" design style of digital electronics engineering gave ground to a "top-down" style, enabling logicians and programmers to build modular systems, with standard boards, for an expanding computer market.

MSI packages, integrating the functions (e.g. 4-bit ALU) of such boards on a single chip, allowed the simpler construction of more complex computers, with even less emphasis on "bottom-up" design. This evolutionary process continued with the availability of LSI packages, integrating major components (e.g. 8-bit microprocessors and 16K RAMs) on single chips, enabling exclusively "top-down" computer designs, based on the selection of a few highly marketed "plug-in" boards, in an increasingly software engineering environment.

Thus the era of VLSI dawned with the semiconductor industry planning even more sophisticated chips (e.g. 16/32-bit microprocessors and 64/256K RAMs) to provide a source for the computer industry which was eagerly awaiting the next generation of component technology. However, this increase in on-chip complexity brought with it a whole new set of problems. Indeed, the new opportunities offered by VLSI were approached very cautiously by the microelectronics industry. A new "learning curve" had to be climbed, and, because of the speculative nature of VLSI chip development, venture capital was difficult to obtain. Consequently, although the problems and opportunities of VLSI were widely discussed, "VLSI chip projects" remained mainly in the LSI camp! Nevertheless, the prospect of more specific VLSI systems remained sufficiently attractive to stimulate research towards cost-effective solutions for the problems.

Clearly, a watershed had been reached in the evolution of integrated circuits, and new design techniques had to be developed. The management of complexity was the main problem, and so the help of computer scientists was enlisted.

Computer science had already faced problems of complexity in the development of large programs, and hence the well-established concepts of modular, hierarchical, block-structured software development were transferred to VLSI design. Accordingly, a trend towards well-structured (viz regular) "VLSI chip architectures" and formal "VLSI design methodologies" was initiated and the development of software "VLSI design tools" was greatly accelerated. Indeed, VLSI design rapidly became an accepted subject area of computer science. Major national programmes were started and new courses, teaching these VLSI disciplines, sprang up all over the industrial world. Indeed, this book results from one such course.

1.1 VLSI Chip Architecture

In contrast to a random collection of transistors, typical VLSI chip layouts (viz floorplans) are partitioned into some ordered configuration of functional blocks, each of which may be composed of a hierarchy of sub-blocks, which are partitioned into some ordered configuration of base blocks. Thus a VLSI the provision of army uniforms; with a "best-fit" or "first-fit" policy, depending on the state of the national economy.

1.1.2 Chip floorplans composition styles

There are five clearly distinguishable styles of chip floorplan composition, which are described below.

1.1.2.1 Place and route. This is a "block-oriented" approach to chip composition, for which two sub-classes are evident.

(*a*) *Random routing*. This style corresponds to the full-custom approach to leaf-cell design, since the leaf cells are placed strategically in the floorplan before interconnection by the shortest possible paths. Consequently, the style leads to random routing patterns, which are time-consuming to layout and test, since they are not amenable to computer assistance. Although, at first thought, such routing seems trivial compared with leaf-cell layouts, it can be very tedious and highly prone to error.

High-volume memory and microprocessor chips, optimized for the highest performance at the lowest cost, usually fall into this category.

(*b*) *Routing channels*. With this approach, linear arrays of leaf cells are separated by routing channels which interconnect at the ends of the arrays. Usually, the leaf cells are constrained to fit the same vertical dimension and are sited to make the best use of the array length. Cells are then interconnected via the appropriate busses in the routing channels. Clearly, this style, in conjunction witht the standard cell style of leaf-cell design, is very well suited to computer generation. Indeed, this combination forms a popular base for commercial "silicon compilers".

Unfortunately, the simplicity of this composition style is offset by the following problems:

(i) resiting of cells is often necessary to ease interconnection bottlenecks;
(ii) poor layout packing density, due to oversize leaf cells and routing channels, leads to cost-ineffective use of silicon;
(iii) different and unknown path lengths of cell interconnections leads to synchronization problems and poor chip performance.

1.1.2.2 Route and slot. This is a "bus-oriented" approach to chip composition, incorporating a much more regular form of routing channel, to ease the problems cited above. Indeed, a uniform routing channel is established on the chip floorplans and then leaf cells are slotted in appropriate locations relative to the channel. Moreover, this style can benefit from the full advantages of computer-aided design. Two sub-classes of route and slot composition are evident.

(*a*) *Standard bus.* Popular for microprocessor chip architectures, this style slots leaf cells on either side of a strategically placed bus. Such busses are usually time-multiplexed to save chip area. In conjunction with the standard cell style of leaf-cell design, this style is well suited to computer generation.

(*b*) *Grid structure.* Currently the most popular routing structure for ULAs and gate arrays, this style is based on an orthogonal grid of routing channels. Leaf cells are slotted in the square "holes" of the grid and connected to the appropriate lines in the channel. Clearly, this style is very well suited to computer generation.

Although considerably easing problems (i), (ii) and (iii) cited above for routing channels, the route and slot style still suffers the following deficiencies:

(i) much of the active chip area is "wasted" on routing channels;
(ii) long signal transmission delays, compared with the logic propagation delays of the leaf cells, degrade overall chip performance.

1.1.2.3 Bit-sliced structures. This is another "bus-oriented" approach to chip composition, which develops the standard bus style described above to minimize its inherent non-functional chip area and signal transmission delays. The composition style is easily recognized, since data and control signals are routed orthogonally on two separate communication layers (e.g. metal and polysilicon in NMOS technology). Leaf cells are bit-sliced and "hung" below a strategically placed data bus, each bit of the cell aligning with the corresponding bit of the bus. Thus, a standard bus is integrated within the leaf cells, which are fixed in height and line-pitch by the dimensions of the bus.

Although this style is not as well suited to computer generation as route and slot compositions, several silicon compilers are being developed for this approach. Two sub-classes of bit-sliced composition are evident.

(*a*) *Abutting cells.* In this style leaf cells can support different logical functions. Cell layouts must conform in height and line-pitch, but cell width (viz along the bus) can be tailored to suit the cell function. Clearly, this approach is well suited to the simplified full-custom style of leaf-cell design. Indeed, the style has found favour with designers of microprocessor data-paths.

Problems can be experienced with this approach when several different cell types are required, owing to the inevitable non-functional chip area suffered by some of the cells.

(*b*) *Iterative cells.* In this style a string of identical leaf cells support specific logic functions with an iterative implementation of a finite state machine. The technique is well known to switching theorists and is applicable to a particular class of logic design; for example parity checkers and serial–parallel multipliers. Usually, more leaf cells are required with iterative solutions, but this is offset by a much better cell packing density.

1.1.2.4 Cellular arrays. This is a "memory-oriented" approach to chip composition, which extends the bit-sliced iterative cell style to an orthogonal array of identical cells. The high regularity of cellular arrays has been widely exploited by LSI RAMs, ROMs, PLAs, CAMs (Content Addressable Memory) and array multipliers. Thus the cellular array constitutes a densely packed fully-engineered standard layout block, which is tailored to support specific logical functions by virtue of its data content. Clearly, cellular arrays are a natural extension of the programmable cell style of leaf-cell design, and consequently they are very well suited to computer generation.

Although achieving high cell packing densities, cellular arrays present formidable circuit design and pin-count problems. Indeed, each cell in all bit columns and each cell in all word rows must be connected to at least one bus, which in turn must be routed to a pin. Thus pin counts could increase as the square root of the number of cells in the array. Since large pin counts suffer high cost penalities and output pad drivers are major contributors to chip power dissipation, these problems of cellular arrays significantly detract from their otherwise distinct advantages.

To ease such input–output problems, cellular arrays often incorporate decoding and multiplexing logic blocks.

1.1.2.5 Distributed logic memories. This is a VLSI generalization of the LSI cellular array aproach to chip composition, in which each cell incorporates memory and processing logic and peripheral logic is included to handle input–output. Thus, logic is distributed over memory in a highly regular and uniform manner. Such chips can support a segment of a data structure, and hence they form ideal building blocks for "fine-grain" distributed computer architectures. Consequently, not only is this style very well suited to the constraints of VLSI, but it is also very well suited to its application environment. Two sub-classes of this composition style are in evidence.

(*a*) *Static configuration.* In this style both the leaf-cell function and intercell communications are dedicated to the support of a particular data structure. Examples include systolic arrays and certain proposals for VLSI string, array and tree processors.

(*b*) *Dynamic configuration.* In this style the leaf-cell and intercell communication network designs are generalized to support programmable cell functions and variable data configuration. For example, a string processor could be programmed to support table processing (by concatenating row entries into a single string) or an array processor could be programmed to support tree processing (by enabling the appropriate intercell connections and inhibiting the others). Indeed, associative processing structures are good examples of this style of chip composition. Moreover, reconfiguration of the communication network allows fault-tolerant VLSI chip architectures to degrade gracefully in the event of cell failure.

1.2 VLSI parallel-processing chip architecture

Progress in the new fields of "VLSI chip architecture", "VLSI design methodology" and "VLSI design tools" brought about a reappraisal of VLSI design complexity problems, and specific VLSI systems projects were initiated. Inevitably, 32-bit microprocessors and 256K dynamic RAMs figured prominently among the first attempts at VLSI chip design. Indeed, more specific VLSI chip architectures, involving memory and processing logic on the same chip, have been proposed. However, VLSI creates the opportunity for the integration of radically new parallel-processing computer architectures, and consequently many VLSI research investigations have taken this direction.

In particular, the "distributed logic memory" style of VLSI chip architecture, described in sub-section 1.1.2.5, has stimulated much research attention for the support of parallel computation.

1.2.1 Parallel computer performance

Various studies, analysing statistical results derived from "benchmark" programs, have distinguished between two types of parallelism, to be exploited by a parallel CPU (viz a Central Processing Unit comprising a number of PEs (Processing Elements)), these being as follows.

(1) *"natural" parallelism:* the potential parallelism of the procedure to be executed, independent of the processing power of the CPU, where

$$\text{potential parallel CPU performance} \equiv P_n = f1\left(\frac{p}{\log_2 p}\right). \quad (1)$$

(2) *"applied" parallelism:* the capability of the parallel CPU to exploit the "natural" parallelism of the procedure to be executed, where

$$\text{actual parallel CPU performance} \equiv P_a = f2(\log_2 p). \qquad (2)$$

The curves for "natural" and "applied" parallelism are shown in Figure 1, clearly illustrating the loss in potential performance suffered by "force-fitting" procedures on a particular non-ideal parallel CPU. The curves demonstrate quite clearly that PE redundancy increases rapidly with increase in the number of PEs; a feature which does not augur well for cost-effectiveness! Indeed, this observation leads to the following recommendations for the design of cost-effective parallel computers.

(1) Reduce the number of PEs to minimize PE redundancy. This is not an attractive option, since it limits the potential increase in CPU processing power, but it is one that is often chosen for general-purpose CPU structures.
(2) Reduce PE cost such that redundant PEs are not cost-significant. Improving technology augurs well for this option. In fact, VLSI has created a considerable opportunity for parallel CPU designers.
(3) Restrict the applications of the parallel CPU to only those with a close-mapping between the "applied" parallelism of the CPU and the "natural" parallelism of the data.

Recommendation (3) suggests that a parallel CPU could be best exploited as a peripheral "add-on" (to a general-purpose computer), which is dedicated

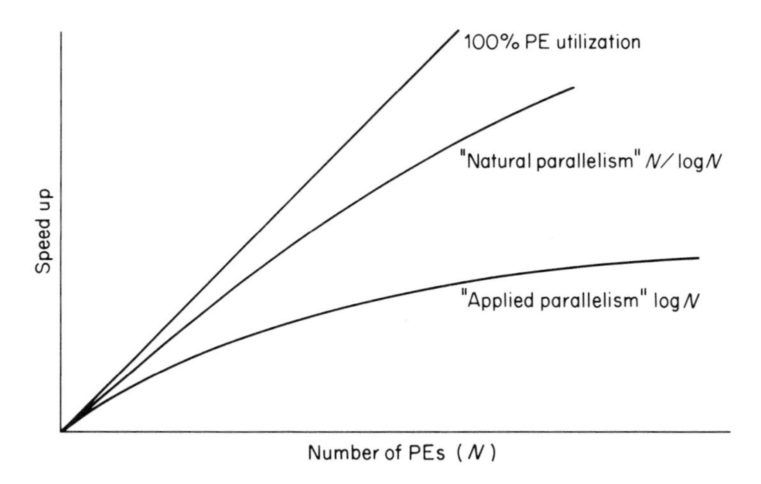

Figure 1 Relative Speed-up With Parallel Processing.

to the execution of specific parallel algorithms. In this computer architecture the sequential CPU would support "general processing" and provide stored-program control of the parallel CPU for "structured data processing". Thus each PE would support a "node" of a specific data structure and the PEs would be interconnected to exactly match the "links" of that data structure.

Consequently, VLSI parallel-processing chip architectures are evolving as flexible building blocks for particular structured data processors (e.g. tree, array and string processors). Clearly, this high replication of simple PEs, with simple interconnections, is very well suited to VLSI, and cost-effective parallel CPU structures are likely to emerge.

Evident among this proliferation of VLSI proposals is the orientation towards either the SIMD (viz array) or the MISD (viz pipeline) parallel CPU structures for specific "non-numerical information processing" (e.g. database management and information retrieval) and "vector and matrix processing" (e.g. digital signal and image processing) applications. VLSI "systolic arrays" and "associative string processors" (as mentioned below) provide practical examples of attempts to integrate such cost-effective parallel-processing CPU structures.

1.2.2 VLSI associative string processors

The author has been investigating microelectronic parallel-processing structures, conforming to the "dyamically configurable distributed logic memory" style of chip architecture (see sub-section 1.1.2.5), for more than a decade. During this time, many architectures for LSI and VLSI fabrication have been investigated. Research objectives can be summarized as the search for a highly programmable architecture which ideally

(1) supports and manipulates structured data (e.g. strings, arrays, tables, trees etc.), and
(2) fits the technological constraints of chip fabrication (e.g. high cell replication, high cell–pin ratio, low power dissipation etc.)

for cost-effective implementation of highly parallel information-processing systems. A noteworthy result of this research has been the concept of the "associative string processor", which has stimulated much industrial interest and is currently the subject of several application studies. Particular variants of this concept will be used as practical examples for the study of VLSI parallel-processing chip architecture in this chapter.

An "associative string processor" comprises a string of identical APEs (Associative Processing Elements), as shown in Figure 2. Each APE is connected to an "inter-APE communication network" (which runs in parallel with the string), data and control busses and a single feedback line. The

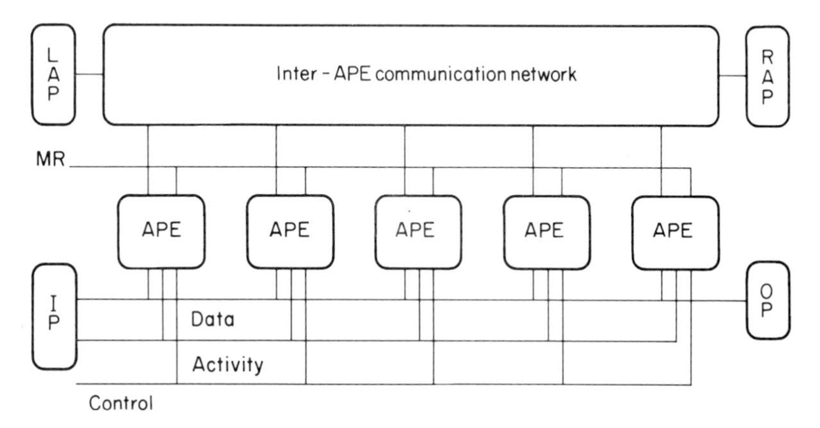

Figure 2 Associative String Processor.

contents of each APE comprise local storage, a comparator and processing and communication logic. In operation, sub-strings are addressed by content for subsequent SIMD parallel processing. Moreover, the string can be segmented for MSIMD parallel processing.

The author has shown that the "associative string processing" concept can be cost-effectively implemented with an "Associative Parallel Processor" (APP), which comprises a "Content Addressable Memory" (CAM) with microprogrammed bit-column and word-row control. CAM chips and a Micro-APP test chip have been fabricated to verify and evaluate the concept. It has been shown that Micro-APP chips can be applied to a wide range of scalar–vector and vector–vector parallel-processing applications. Currently, two Micro-APP chip variants are in development; the SCRIPT (Single Chip Relational Information Processing elemenT) chip and SCAPE (Single Chip Array Processing Element) chip for text and image processing respectively. These VLSI chips will be used as practical examples of VLSI parallel-processing chip architecture in the remainder of this chapter.

2 NON-NUMERICAL INFORMATION PROCESSING

An increasing proportion of the workload of modern computer systems involves the representaton and manipulation of mainly non-numerical data structures. For example, editing, compiling, file-processing, table sorting and searching all involve extensive processing of symbolic data which have no numerical meaning. Moreover, the "information explosion", seen in recent years in all fields of human endeavour, has already stimulated the development of computer-based "information systems" to assist the creation, modification, classification, storage, retrieval and dissemination of mainly "textual" data. Furthermore, the current interest in "database management systems", "expert systems" and "intelligent knowledge-based systems" will inevitably increase the demand for non-numerical information processing. Nevertheless, minicomputer and microcomputer hardware is almost invariably designed for mainly numerical computation. Clearly, design optimization for fast high-precision arithmetic provides little benefit for non-numerical information processing. Indeed, the lack of flexible data-structuring facilities is a major impediment to cost-effective information-system design.

Fortunately, there are signs that the future for non-numerical information-processing systems is rather more encouraging; the increasing cost of software development is already seriously limiting its cost-effectiveness over an increasingly wide range of computer applications. Moreover, at the same time, the rapidly growing interest in "information technology" and the prospect of VLSI (Very Large Scale Integration) are stimulating proposals for radically new microcomputer chip architectures. In particular, VLSI offers the potential of low-cost building blocks for the flexible support and manipulation of non-numerical data structures.

2.1 Associative String Processing

A common feature of non-numerical information-processing applications is the requirement for a "set" of data elements, "associated" with a common "key" (or Boolean combination of "search keys"), to be referenced, by virtue of the "value" of the key(s) and independently of its location within a data structure, for the purposes of retrieval or update. To illustrate the nature of such "associative processing", consider the following examples of retrieval and update with the telephone directory shown in Table 1.

Example 1. The telephone number of the person named Scott in the SAles department can be found with the following "associative process":

```
forall i     : (Name[i] = Scott) and (Dept[i] = SA) do
  Number := Num[i]
```

Example 2. The telephone numbers of all ACcounts personnel in room 127 can be changed to 134 with the following "associative process":

```
forall i   : (Dept[i] = AC) and (Room[i] = 127) do
  Num[i] := 134
```

Table 1 Segment of an internal telephone directory.

Name	Style	Dept	Room	Num
Scholes	Mr	AC	127	137
Scorer	Dr	RE	64	109
Scott	Mrs	SA	201	193
Scowen	Miss	AC	127	137
Scrips	Mr	SA	202	194
Scriven	Dr	RE	65	110
Scroggs	Mrs	AC	127	137
Scroll	Dr	RE	64	109
Scrow	Mr	AC	128	135

Clearly, "associative processing" is applicable to a wide range of such relational data processing operations (e.g. intersection, union, difference, Cartesian product and join operation on two relations and projection, selection and mapping on one relation).

A particularly flexible form of "associative processing" is that of "associative string processing", since all data structures can be mapped to string format (e.g. structured data stored as a disk file). For example, the entries of a data table can be simply concatenated to form a long string, with the added advantage that variable-length entries can be easily accommodated. In fact any linked-list structure (e.g. queues, rings, trees etc.) could be represented in string format, by concatenating records of the type

```
 -------------------------------------
| in-links  |  value  |  out-links  |
 -------------------------------------
```

In summary, "associative string processing" can be regarded as an intermediate processing level, common to a wide range of non-numerical information-processing tasks.

2.2 Associative String Processing: SISD Implementation

As an illustration of the inadequacy of the SISD computer architecture for "associative string processing", consider its support of "key searching".

The following algorithm displays the first character following the first sub-string of a text string (represented as the text-file S) matching the N-character "search-key" (represented as the character-array K).

```
j := 1;                  {key-address pointer initialization}
while not eof (S) and (j <= N) do
  begin
    read (S,ch);         {string scanning}
    if K[j] = ch         {character matching}
      then j := j + 1    {key-address pointer incrementing}
      else j := 1        {key-address pointer back-tracking}
  end;
if not eof (S)
  then begin
         read (S,ch);
         writeln (ch)
       end
  else writeln ('No match found')
```

The algorithm clearly demonstrates two forms of delay.

(*1*) *Latency delay:* the time taken to scan the characters that precede the first matching sub-string. Latency delay is highly application-dependent (since a matching sub-string may be found anywhere or, indeed, nowhere in the string), and hence this delay is limited by string formatting (viz file organization) and string data rate. In practice, much effort (e.g. file-sorting) is expended to reduce latency delay. For example, assuming that the text string is segmented into data "records", Figure 3 clearly shows that latency delay is critically dependent on the ordering of the "records" within the string.

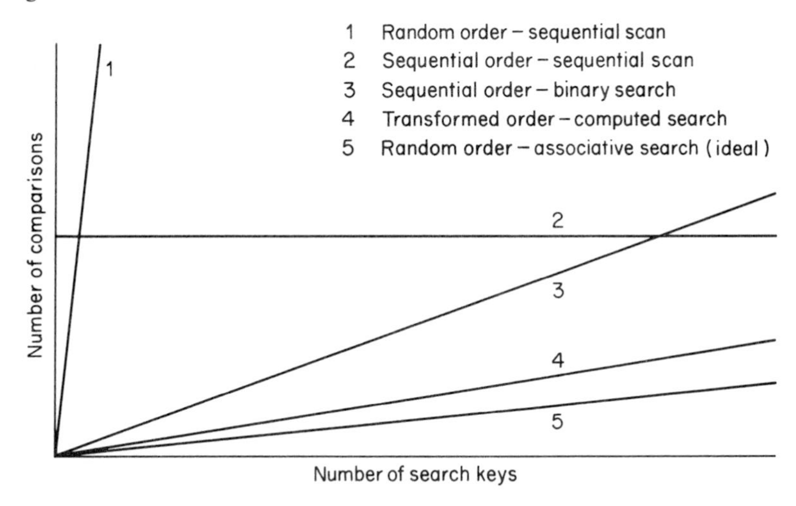

Figure 3 Key-Searching Comparison.

(2) *Match delay:* the time taken to compare the sub-string with the "search key". Match delay is less application dependent (since a "search key" is short compared with string length), but this delay is limited by the complexity of the string-processing operation, key storage delays, character comparison, key-address pointer incrementing and back-tracking delays as well as string data rate.

For a single "search-key", match delay is negligible compared with latency delay. However, in practice, "queries" usually comprise a Boolean combination of "search keys" and include more complicated string-processing operations than simple matching (e.g. partial and threshold matching). In such cases, key-address pointer following is considerably complicated, since text sub-strings must be compared with each "search key" before the Boolean condition can be tested.

The curves show that the "associative file searching" efficiency of SISD computer hardware can be improved, but at the expense of extra software or firmware complexity (e.g. with sorting, indexing or hashing techniques). Indeed, for more general "associative string processing" this additional complexity can lead to unpredictable, inflexible and costly information-processing systems.

2.3 Associative String Processing: SIMD "Associative Parallel Processor" Implementation

The Associative Parallel Processor (APP) is ideally matched to the execution of the "associative processes" described in sub-section 2.1, and hence its implementation has been of interest for many years. Unfortunately, the APP is notoriously difficult to integrate, and, up till now, no truly cost-effective implementation has yet been announced. Thus the possibility of a VLSI APP chip is an exciting research prospect.

As mentioned in sub-section 1.2.2, the author has proposed several Micro-APP variants [1–3] for cost-effective "associative string processing", and, of these, the SCRIPT chip is optimized for text-processing applications.

2.3.1 The SCRIPT chip

The SCRIPT (Single-Chip Relational Information Processing elemenT) chip executes "associative processes" (viz "search" and "read" or "write") under the control of an external "microprogram sequencer". In fact, the SCRIPT chip floorplan comprises a close-packed structure of four different

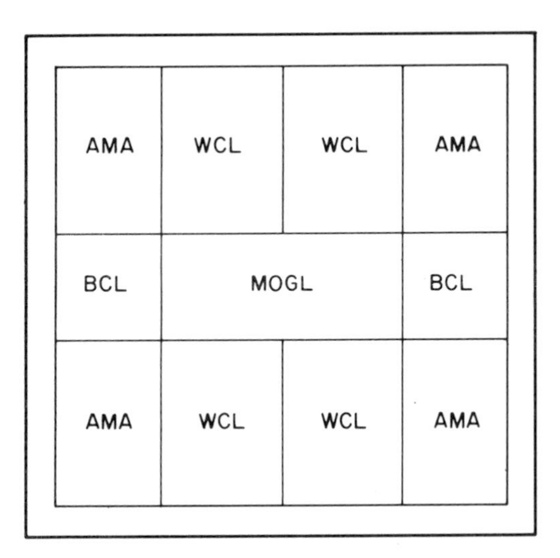

Figure 4 SCRIPT and SCAPE Chip Floorplan.

(but exactly butting) functional blocks, as shown in Figure 4, namely the following.

BCL: Bit Control Logic selecting 0 or more of the 40 AMA bit columns in support of a bit-parallel "search" or "write" operations on four 8-bit character fields and an 8-bit control field within the AMA. The BCL also includes a 40-bit Data Input Register for "search" and "write" operations and a 32-bit Data Output Register for "read" operations.

AMA: Associative Memory Array, a Content Addressable Memory (CAM) comprising 256 40-bit word rows. During a "search" operation, each AMA bit cell compares its content with the corresponding bit in the Data Input Register and generates the appropriate match or mismatch signal; the bit cells in each masked bit-column generate match signals. If no bit cell in a particular word row generates a mismatch signal, then that word row is deemed to have matched the "search" word. A "read" operation transfers the contents of an activated word-row to the Data Output Register, whereas a "write" operation updates selected fields of activated word rows with the unmasked contents of the Data Input Register.

WCL: Word Control Logic activating 0 or more of the 256 AMA word rows, for "read" and "write" operations, according to a defined mapping on the response to a "search" operation, which is stored in one of two Tag Registers. Activation mappings include the resolution of the first (last) matching word row, neighbour and group activation and linking between remote matching AMA word rows. The WCL also includes the Match Reply, MR, a 1-bit flag which indicates the presence of one or more tags in the Tag Register.

MOGL: Micro-Order Generation Logic issuing "dynamic micro-orders" to the BCL and WCL derived from the "static micro-order" of the current microinstruction. Input–output multiplexing (to minimize pin count and package dimensions) and internal clock generation are also performed in the MOGL.

SCRIPT chips would be linked to form a "SCRIPT chain", and the "SCRIPT chain controller" would support a set of "SCRIPT macros". To illustrate the application of a "SCRIPT module" (viz a "SCRIPT chain" and a "SCRIPT chain controller" on a single printed-circuit board) to general text string processing, the following Pascal procedure finds and marks all occurrences of the sub-string 'PROCESS' in the "SCRIPT chain".

```
procedure FIND_SUB_STRING (ss:string);
var i : 1..n; {assuming sub-string length of n}
begin {FIND_SUB_STRING}
  INITIALIZE;
  i := 1;
  MARK_HEADS (ss[i]);
  repeat
    i := i + 1;
    FOLLOW_TAILS (ss[i])
  until (i = n) or (MR = mismatch);
  if MR = mismatch
    then writeln ('Sub-string ',ss,' not found')
    else ISOLATE_SUB_STRINGS
end {FIND_SUB_STRING}
```

The effects of the "SCRIPT macros" INITIALIZE, MARK_HEADS, FOLLOW_TAILS and ISOLATE_SUB_STRINGS are illustrated in Tables 2, 3, 4, 5 and 6 respectively.

Table 2 State of the “SCRIPT chain” after INITIALIZE.

Search data	Write data		Content of the Associative Memory Array																					
X	0	B4																						
X	0	B3																						
X	0	B2																						
X	0	B1																						
*	*	Ch	S	T	R	I	N	G	P	R	O	C	E	S	S	I	N	G	P	R	O	G	R	A	M

Table 3 State of the “SCRIPT chain” after MARK_HEADS (ss[1]).

Search data	Write data		Content of the Associative Memory Array																					
X	X	B4																						
X	X	B3																						
X	2	B2								2										2					
1	X	B1							1										1						
P	*	Ch	S	T	R	I	N	G	P	R	O	C	E	S	S	I	N	G	P	R	O	G	R	A	M

Table 4 State of the “SCRIPT chain” during FOLLOW_TAILS (ss[i]).

Search data	Write data		Content of the Associative Memory Array																					
X	X	B4																						
X	X	B3																						
X	X	B2																						
R,2	2	B2									2										2				
O,2	2	B2										2										2			
C,2	2	B2											2												
E,2	2	B2												2											
S,2	2	B2													2										
2	2	B2														2									
X	X	B1							1										1						
S	*	Ch	S	T	R	I	N	G	P	R	O	C	E	S	S	I	N	G	P	R	O	G	R	A	M

Table 5 State of the "SCRIPT chain" during the execution of ISOLATE_SUB_STRINGS.

Search data	Write data		Content of the Associative Memory Array
X	X	B4	
X	3	B3	3 3 3 3 3 3 3
2	X	B2	2
X	X	B1	1 1
*	*	Ch	STRING PROCESSING PROGRAM

Table 6 State of the "SCRIPT chain" after ISOLATE_SUB_STRINGS.

Search data	Write data		Content of the Associative Memory Array
X	X	B4	
3	X	B3	
X	X	B2	2
1	1	B1	1
*	*	Ch	STRING PROCESSING PROGRAM

The SCRIPT chip corresponds to the "dynamic configuration distributed logic memory" style of VLSI chip architecture defined in sub-section 1.1.2.5. As such, it can be programmed to support general "associative string processing", and consequently a wide range of non-numerical information-processing systems could be based on the SCRIPT chip. The chip, supporting 100 ns "associative processes" and comprising around 150K transistors on a 100-pad CMOS die, is due to be fabricated by Plessey in 1986.

3 DIGITAL SIGNAL PROCESSING

Digital signal-processing applications (e.g. filtering, correlation and Fourier transformation) involve a great deal of vector and matrix arithmetic, which is amenable to execution with parallel CPU structures.

As an example, consider the execution of "matrix multiplication", assuming the type definition

A,B,C = array [1..n,1..n] of integer

and the "scalar-processing" algorithm

```
for i := 1 to n do
  for j := 1 to n do
    begin
      C[i,j] := 0;
      for k := 1 to n do
        C[i,j] := C[i,j] + A[i,k] * B[k,j]
    end
```

Thus matrix multiplication involves the execution of three nested loops, namely 1 "outer product" (viz the i-loop), n "middle products" (viz the j-loop) and n^2 "inner products" (viz the k-loop). Note that for "scalar processing" the computational complexity of the "inner-product" calculation is n multiply–add operations and the matrix multiplication time is $O(n^3)$.

At first sight, it would appear that "matrix multiplication" is fundamentally a "sequential process", since the "inner product" is calculated as a sequence of single-valued multiply–add operations. However, further consideration reveals that by re-ordering the loops, two different "vector-processing" algorithms can be found; these are

```
(1) for j := 1 to n do
      for k := 1 to n do
        forall i do
          C[i,j] := C[i,j] + A[i,k] * B[k,j]

(2) for i := 1 to n do
      for k := 1 to n do
        forall j do
          C[i,j] := C[i,j] + A[i,k] * B[k,j]
```

In each case, the "inner product" is calculated in parallel (for all elements in the ith row (algorithm 2) or jth column (algorithm 1) of matrix C) in a two-step process; that is a "scalar–vector multiplication" followed by a "vector–vector" addition. Thus for "vector processing" the computational complexity of the "inner-product" calculation is one parallel multiply–add operation and the matrix multiplication time is $O(n^2)$.

The choice of "vector-processing" algorithm (1) or (2) depends on how storage is allocated to the elements of A, B and C, relative to its input–output data ports. In general, if memory is allocated to provide parallel access to the elements of a matrix row then the columns of that matrix can only be accessed sequentially. For example, in algorithm (1), the elements of the *j*th column of *C* and the elements of the *k*th column of *A* must each have access to an independent memory port. However, by allocating store locations along a common diagonal (viz "skew storage") to a matrix row, parallel access can be gained to both its rows and columns; whereupon the selection of algorithm (1) or (2) becomes arbitrary.

Further consideration of the "matrix-multiplication" algorithm leads to the realization that the "middle product" can be also calculated in parallel (for all elements of *C*) in a two-step process; that is with a "vector–vector multiplication" followed by a "vector–vector" addition, as follows:

```
for k := 1 to n do
   forall i,j do
      C[i,j] := C[i,j] + Ad[i,k] * Bd[k,j]
```

where *Ad* and *Bd* represent the replications of the *k*th column of *A* and the *k*th row of *B* respectively. Thus for "matrix processing" the computational complexity of the "middle-product" calculation is one parallel multiply–add operation and the matrix multiplication time is $O(n)$.

Two prominent and related examples of digital signal processing are those of "linear filtering" (e.g. in telecommunications applications) and "spatial filtering" (e.g. in image-processing applications).

3.1 Linear Filtering

A basic building block for the construction of linear filters is shown in Figure 5, and its response is described by the linear convolution sum, as given below:

$$Y_i = \sum_{m=0}^{N} W_m X_{i-m}$$

where

N is the order of the filter block,
X_i is the *i*th input sample,
W_m is the *m*th filter weight,
Y_i is the *i*th output sample.

Actual FIR (Finite Impulse Responses) and IIR (Infinite Impulse Response) filters can be built with "cascade" and "parallel" configurations [4] of such filter blocks and "tuned" by selection of the appropriate weight values.

3.1.1 *Linear filter: SISD implementation*

Current implementations of filter blocks usually employ a bit-sliced micro-processor to control a high-speed multiplier–accumulator chip (e.g. TRW

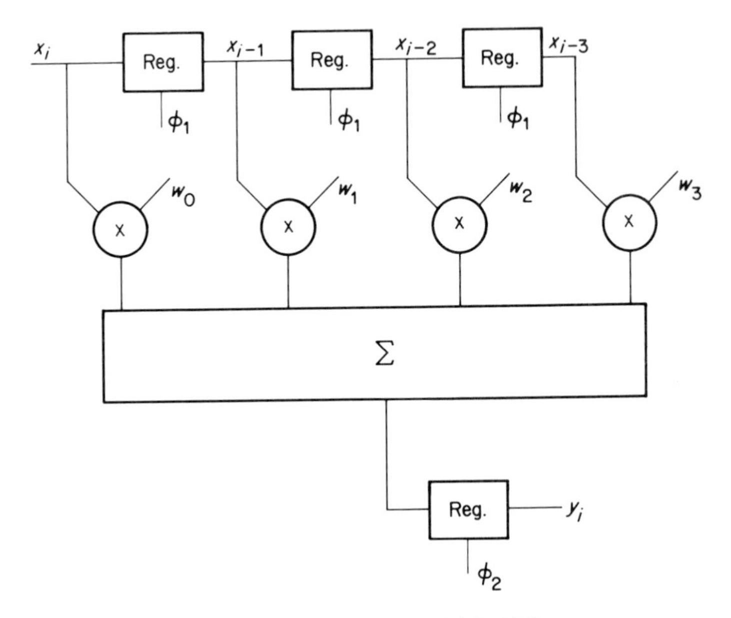

Figure 5 Third-Order FIR Digital Filter.

MPY-16) or a special-purpose microprocessor (e.g. TI TMS 320) incorporating a fast multiplier, with a sequential algorithm of the form

```
S := 0;
for m := 0 to N do
  S := S + W[m] * X[i–m];
Y[i] := S
```

where S is the sum held in the accumulator. The performance of such sequential processing implementations can be summarized as follows:

$$\text{sample rate} = \frac{\phi}{N+1}, \tag{3}$$

$$\text{phase delay} = \frac{N+1}{\phi}, \tag{4}$$

where ϕ is derived from "register load" (T_1), "multiply" (T_m) and "add" (T_a) delays as follows:

$$\phi = \frac{1}{T_1 + T_m + T_a}. \tag{5}$$

Since each sample requires the execution of $N + 1$ steps to complete the inner-product calculation for each filter block, high-speed signal processing can only be achieved with very high multiply–add rates. Moreover, high-speed stores and register blocks are required for sample, weight and intermediate result storage. Such high-speed components lead to high-cost filter implementations. Nevertheless, the data transfer "bottleneck" between store and register block constitutes a major limitation to processing speed.

3.1.2 *Linear filter: bit-serial SIMD implementation*

The practical limitations of the SISD implementation have led the author [5] to suggest a lower-performance, but more cost-effective, bit-serial SIMD chip architecture for the implementation of linear filters. This approach is based on the "dynamic configuration distributed logic memory" style of VLSI chip architecture, defined in sub-section 1.1.2.5. As shown in Figure 6, the chip floorplan is dominated by a RAM which supports $N + 1$ words of

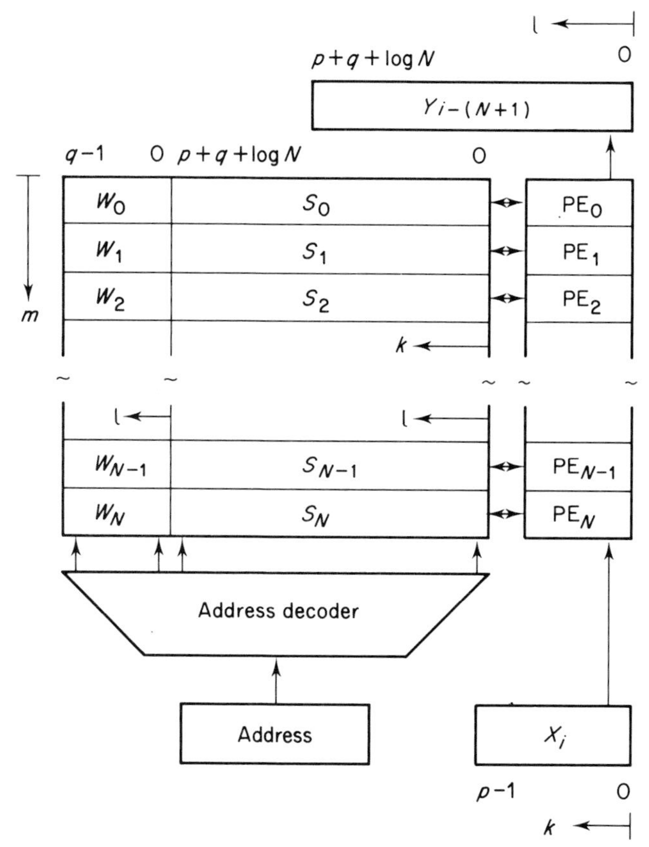

Figure 6 Bit-Serial SIMD Digital Filter.

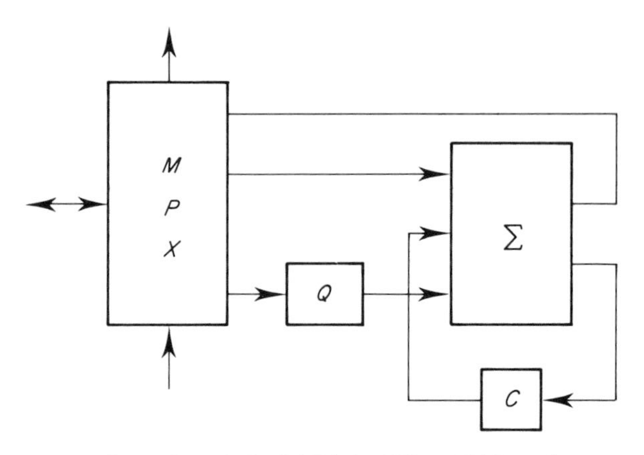

Figure 7 Bit-Serial Digital Filter PE Detail.

$p + 2q + \log_2 N$ bits. Each word has independent access to a dedicated PE, as shown in Figure 7, such that all $N + 1$ bits of addressed bit slices can be presented for "bit-serial word-parallel" processing. Hence this highly regular and very compact structure can be "programmed" to support linear filtering with an algorithm of the form

```
forall m do S[m] := S[m] + W[m] * X[i];     {Multiply-add}
forall m do S[m] := S[m+1];                 {Shift}
Y[i-N-1] := S[0]
```

where the "multiply-add" is performed bit-serially as follows:

```
for k := 0 to p-1 do
  if X[i,k] = 1
    then for l := 0 to q-1 do
          S[m,k+l] := S[m,k+l] + W[m,l]
```

and the "shift" could be performed bit-parallelly within the RAM structure, with the following sample rate

$$\text{sample rate} = \phi = \frac{1}{pq(2T_r + T_a + T_w) + T_r + T_s + T_w}; \qquad (6)$$

or bit-serially as follows:

```
for l := 0 to p+q-1+log2 N do
  S(m,l] := S[m+1,1]
```

with the following sample rate:

$$\text{sample rate} = \phi = \frac{1}{pq(2T_r + T_a + T_w) + (p + q + \log_2 N)(T_r + T_s + T_w)} \tag{7}$$

and the following phase delay

$$\text{phase delay} = \frac{N + 1}{\phi} \tag{8}$$

where ϕ is derived from the "read" (T_r), "write" (T_w), "add" (T_a) and "shift" (T_s) delays, as shown above.

Since this implementation operates bit-serially, variable filter "orders" and "sample" and "weight" precisions can be easily accommodated.

3.1.3 Linear filter: comparative performance evaluation

In order to compare the relative merits of the above mentioned implementations, the following linear filtering environment will be assumed:

order of the filter = 9
sample precision = 12 bits
weight precision = 8 bits
system clock rate = 10 MHz

For the SISD implementation it is assumed that register loading and adding require one clock cycle each and that muliplication will require two cycles. For the bit-serial SIMD implementation, it is assumed that the chips would support a 40 MHz internal clock and that each operation requires one internal clock cycle.

The data shown in Table 7 indicate that the bit-serial SIMD implementation, although slower (for a 9th-order linear filter) than the SISD approach, integrates the filter on a single VLSI chip. Whereas the SISD implementation would require two chips; namely a RAM chip and a multiplier–accumulator chip. Moreover, the control algorithm for the bit-serial SIMD implementation is so simple that it could be incorporated on the same chip,

Table 7 Comparison of 9th-order linear filter implementations.

Implementation	Maximum sample rate	Minimum phase delay
SISD	0.25 MHz	4 μs
bit-serial SIMD		
bit-parallel shift	0.10 MHz	97 μs
bit-serial shift	0.08 MHz	120 μs

thereby simplifying the requirement for external control logic. On the other hand the SISD implementation requires additional control circuitry.

Significantly, the performance of the SISD implementation is dependent on the order of the filter, whereas the performance of the bit-serial SIMD implementation is dependent on sample and weight precision.

Overall, the bit-serial SIMD approach is likely to be the more cost-effective, especially for higher-order linear filters requiring lower bit precision.

The performance of both implementation styles could be improved by design optimization. However, this may be cost-effective for only the SISD implementation, since its components can be marketed in a much wider application range.

3.1.4 Linear filter: associative string processor implementation

The VLSI "associative string processor" concept (discussed in sub-section 1.2.2) offers a more flexible possibility than the bit-serial SIMD implementation, for linear filtering. Executing similar algorithms, an "associative string processor" could support a "cascade" or "parallel" array of FIR filter stages, as a string of concatenated "weight" (entries distinguishable by content address), within a single chip. In this case, the chip is likely to outperform SISD implementations, by exploiting a higher degree of parallelism. Being a programmable structure, a VLSI "associative string processor" could support many different scalar–vector and vector–vector parallel-processing algorithms. Thus such VLSI "associative string processor" chips could be thought of as "universal signal-processing elements".

3.2 Spatial Filtering

The convolution of a $K \times K$ kernel (or "weight map") with an input image X to produce an output image Y is given by the following expression

$$Y_{i,j} = \sum_{m=i-k}^{i+k} \sum_{n=j-k}^{j+k} W_{m,n} X_{i-m,j-n}$$

where $k = \frac{1}{2}(K-1)$, k is odd, and

$X_{i,j}$ is the luminance value at the input image location (i, j),
$W_{m,n}$ is the weight value at the kernel location (m, n),
$Y_{i,j}$ is the luminance value at the output image location (i, j).

3.2.1 Spatial filter: SISD implementation

Assuming a SISD implementation and a $Mp \times Np$ "patch", with the origin

(Oi, Oj), within an $M \times N$ image, spatial convolution can be achieved with a sequential algorithm of the form

```
for i := Oi to Oi+Mp-1 do
  for j := Oj to Oj+Np-1 do
    begin
      S := 0;
      for m := -k to k do
        for n := -k to k do
          S := S + W[m,n] * X[i-m,j-n];
      Y[i,j] := S
  end
```

where S is the sum held in the accumulator.

The performance of this SISD implementation of spatial convolution can be summarized as follows:

$$\text{time for } K \times K \text{ convolution over a } Mp \times Np \text{ "patch"} = \frac{Mp\ Np\ K^2}{\phi} \tag{9}$$

where ϕ is as defined in equation (5).

Assuming a 10 MHz clock rate, for a high-speed "state-of-the-art" SISD processor, Table 8 evaluates equation (9) for typical convolution "window" sizes.

Significantly, equation (9) does not allow for "patch" load and dump operations. Hence, load/dump times provide an interesting reference for the data of Table 8. Assuming a 10 MHz pixel rate, the corresponding "patch" and "image" load/dump times are 0.4 ms and 26.2 ms respectively.

Clearly, SISD processors are not capable of achieving the 40 ms target of real-time image processing systems (i.e. based on a two-field interlaced raster-scan TV with a 50 Hz mains frequency) for even a single spatial convolution. Indeed, when more than one algorithm must be executed within this time constraint, the SISD processor is woefully inadequate.

Table 8 Spatial convolution times for an SISD processor.

$K \times K$	64×64 patch	512×512 image
3×3	3.7 ms	236 ms
5×5	10.2 ms	655 ms
7×7	20.1 ms	1285 ms
$\phi = 10$ MHz		

3.2.2 Spatial filter: bit-serial SIMD implementation

Attempts to achieve higher performance for more flexible real-time image processing have led to many proposals based on bit-serial SIMD processors for this application. In fact, this field of parallel-processing computer architecture is already well advanced; with major contributions including Unger's machine [6], Westinghouse's Solomon computer [7], UCL's CLIP [8], ICL's DAP [9] and Goodyear's MPP [10]. A significant feature of such proposals is the development of special-purpose parallel-processing LSI chips; for example, custom-designed 8-pixel nMOS chips for the CLIP and MPP projects and a 16-pixel gate array for the DAP project. More recently GEC's 32-pixel GRID chip [11] and the author's 256-pixel SCAPE chip (see sub-section 3.3) are exploring the potential of VLSI CMOS for this application. Such chips are based on the "dynamic configuration distributed logic memory" style of VLSI chip architecture defined in sub-section 1.1.2.5.

Assuming a bit-serial SIMD implementation and a $Mp \times Np$ "patch" with the origin (Oi, Oj), spatial convolution can be achieved with a parallel processing algorithm of the form

```
forall i,j do Y[i,j] := 0;
for m := −k to k do
   for n := −k to k do
      forall i,j do Y[i,j] := Y[i,j] + W[m,n] * X[i−m,j−n]
```

where the "multiply-add" is performed bit-serially as follows

```
for b := 0 to p−1 do
   if W[m,n,b] = 1
      then for l := 0 to q−1 do
               Y[i,j,b+l] := Y[i,j,b+l] + X[i−m,j−n,l]
```

for p-bit pixels and q-bit weights.

The performance of this bit-serial SIMD implementation of spatial convolution can be summarized as follows

$$\text{time for } K \times K \text{ convolution over a } Mp \times Np \text{ "patch"} = \frac{pqK^2}{\phi} + T_s \quad (10)$$

where ϕ is defined as

$$\phi = \frac{1}{2T_r + T_a + T_w} \quad (11)$$

and T_s represents shifting delays for non-neighbour communication.

Assuming 8-bit pixels and weights and a 10 MHz clock rate, for a typical

Table 9 Spatial convolution times for a bit-serial SISD processor.

$K \times K$	64×64 "patch"	512×512 "sub-image"
3×3	230 μs	14.7 ms
5×5	654 μs	41.9 ms
7×7	1310 μs	83.8 ms
ϕ = 10 MHz		

bit-serial SIMD processor, Table 9 evaluates equation (10) for common "window" sizes.

Comparing the data of Table 9 with those of Table 8 reveals a 16 times improvement in performance, bought at the expense of many extra chips. For example 512 Goodyear MPP chips, 256 ICL DAP chips or 128 GEC GRID chips would be required for the implementation of the 64×64 pixel array. In practice, design optimization has improved the performance of these chips by a factor of 2–4. Nevertheless, the question of cost-effectiveness still remains.

Although bit-serial SIMD implementations offer the potential of more than one "sub-image process" per frame time, the overheads of "patch" load and dump are no longer negligible. Indeed, a 10 MHz pixel rate would support 64×64 "patch" and "sub-image" load/dump times of 0.4 ms and 26.2 ms respectively. Consequently, such image processors often incorporate local storage and multipixel input–output channels to minimize the time penalty of "patch" load and dump.

3.3 The SCAPE Chip

The SCAPE (Single-Chip Array Processing Element) chip [12] is a variant of the VLSI "associative string processor" concept, which has been optimized for the cost-effective execution of a wide range of real-time linear and, especially, spatial signal-processing algorithms. In fact, the SCAPE chip combines the bit-serial SIMD features discussed in sub-sections 3.1.4 and 3.2.2.

The SCAPE chip floorplan comprises a close-packed square structure of four different (but exactly "butting") functional blocks, as shown in Figure 4; these are as follows.

AMA: Associative Memory Array comprising 256 words, supporting a 32-bit "data field" and a 5-bit "control field", partitioned into 32

8-word content-addressable memory blocks, each block corresponding to an 8-pixel "row segment".

BCL: Bit Control Logic selecting 0 or more of the 37 AMA bit columns in support of arithmetic, logical and relational processing on "declared fields" of activated AMA words. The BCL incorporates index control logic for bit-serial operations and field masking logic. In addition, the BCL also supports a bit-parallel capability, to allow 8-bit and 32-bit processing of activated AMA words.

WCL: Word Control Logic activating 0 or more of the 256 AMA words, according to a defined mapping on the response to a content search of selected AMA bit columns. The WCL incorporates 256 1-bit adder/complementer stages to support bit-serial addition and subtraction operations. Two 256-bit "tag" registers, one to "tag" the currently matching word rows and the other to "tag" previously matching word rows, are also included. Activation mappings control the inter-pixel connectivity, required by image processing algorithms, according to the content of the "tag" registers. The WCL is partitioned into 8-word blocks, one for each of the 32 8-pixel "row segments" in the AMA. These partitions are interconnected by "segment links", which can be set to configure the "row size" (viz 0, 8, 16, 32, 64, 128, 256 or more pixels or, indeed, any multiple of 8 pixels) supported by the SCAPE chip. "Row links" can be "opened" or "closed" such that rows can be processed in "parallel" or "linked" mode. The "chip links" at the ends of the WCL are used to link the SCAPE chips forming the "SCAPE chain".

MOGL: Micro-Order Generation Logic issuing "dynamic micro-orders" to the BCL and WCL derived from the "static micro-order" of the current microinstruction. Input–output multiplexing (to minimize pin count and package dimensions) and internal clock generation are also performed in the MOGL.

For linear signal processing, sample and weight vectors can be stored in the SCAPE chip as a string of integer or real values; with one or more AMA words being allocated to each value. However, for spatial signal processing, two-dimensional pixel and weight vectors must be mapped to the one-dimensional string, by concatenating image rows.

An $n \times m$ image of p-bit pixels can be processed in a singly linked string of n/r SCAPE chips known as the "SCAPE chain"; each SCAPE chip supporting 256 pixels from r rows of the patch.

Table 10

No. of pixels per chip, $r \times m$	No. of pixels per row, m	No. of rows per chip, r	Patch size	No. of chips
256	8	32	16 × 16	1
	16	16	32 × 32	4
	32	8	64 × 64	16
	64	4	128 × 128	64
	128	2	256 × 256	256
	256	1		

The value of p can be selected, under program control, according to the need of the application, with no processing restrictions until p exceeds 8 bits. However, although under program control, the values of n and r affect the complexity of the "SCAPE chain", as shown in Table 10.

The SCAPE chip, comprising 143K transistors on a 68-pin CMOS die, will be fabricated by Plessey (Caswell), and first samples are expected in the second quarter of 1986. Hence, until that time, no accurate data on performance can be given. However, assuming a SCAPE module comprising a 64 × 64 8-bit pixel array (i.e. 16 SCAPE chips), and a 10 MHz clock rate, the following forecasts give some idea of SCAPE performance:

add/subtract 8-bit fields	1.00 μs
scalar–vector 8-bit multiplication	5.45 μs
vector-vector 8-bit multiplication	11.60 μs
3 × 3 spatial convolution (8-bit weights)	95.30 μs
ϕ = 10 MHz	

To reduce pixel input–output overheads, the SCAPE chip includes bit-parallel loading and dumping circuitry which can exchange all 256 pixels within 6.4 μs.

REFERENCES

[1] R.M. Lea, Associative processing of non-numerical information. In *Computer Architecture* (ed. G. G. Boulaye and D. W. Lewin), pp. 171–215 (D. Reidel, Dordrecht, 1977).

[2] R.M. Lea, I^2L micro-associative processors. *ESSCIRC 79 Dig. of Tech. Papers*, pp. 104–106 (1979).

[3] R.M. Lea, Associative processing In *Advanced Digital Systems* (ed. I. Aleksander), Chap. 10 (Prentice-Hall, London, 1984).

[4] J.G.M. Goncalves and R.M. Lea, LSI module for the implementation of digital filters. *Proc. IEE*, Pt F, **128**, 353–358.

[5] R.M. Lea, A VLSI chip architecture for real-time digital filters. *Brunel University Tech. Memo.* CM/R/127 (May 1982).
[6] S.H. Unger, A computer oriented towards spatial problems. *Proc. WJCC*, pp. 234–239 (1958).
[7] D.L. Slotnick, W.C. Borck and R.C. McReynolds, The SOLOMON computer. *Proc. AFIPS* (*FJCC*), pp. 97–107 (1962).
[8] M.J.B. Duff, Review of CLIP-4 image processing systems. *Proc. NCC*, pp. 1055–1060 (1978).
[9] P.M. Flanders, D.J. Hunt, S.F. Reddaway and D. Parkinson, Efficient high-speed computing with the Distributed Array Processor. In *High Speed Computer and Algorithm Organisation*, pp. 113–127 (Academic Press, London, 1977).
[10] K.E. Batcher, Design of a Massively Parallel Processor. *IEEE Trans. Comput.* **29**, 836–840 (1980).
[11] M.M. McCabe, A.P.H. McCabe, B. Arambepola, I.N. Robinson and A.G. Corry, New algorithms and architectures for VLSI. *GEC J. Sci. Tech.* **48**, pp. 68–75 (1982).
[12] R.M. Lea, SCAPE: A Single Chip Array Processing Element for image analysis. In *VLSI 83 Trondheim, VLSI Design of Digital Systems* (ed. F. Anceau and E.J. Aas), pp. 285–294 (Elsevier/North-Holland, Amsterdam, 1983).

2. VLSI complexity, efficient VLSI algorithms and the HILL design system

Th. LENGAUER and K. MEHLHORN

1 INTRODUCTION

In this paper we discuss the relation of algorithms and VLSI in two ways: algorithms for VLSI design and algorithms implemented in VLSI.

In Section 4 we shall discuss the main parts of the HILL design system which is under development at the University of Saarbrücken. The main parts are the layout specification language, the compacter and the switch-level simulator. The HILL layout language provides a convenient way of describing a layout symbolically either by a HILL program or by an interactive graphics session. A HILL program describes a layout at the level of stick diagrams enhanced by extensive means of structuring a design hierarchically. The level of stick diagrams was used previously in systems like CABBAGE, STICKS and MULGA. In contrast with HILL, these systems are graphics-oriented and can therefore support only limited mechanisms for structuring the design, mainly composition and simple iteration. In contrast with that, HILL offers the full power of a high-level programming language (HILL is a PASCAL extension); in particular it offers recursion, iteration and fully parameterized designs. Recursion and iteration are central to (software) algorithm design and we believe that they will be equally important for hardware design. We give some concrete examples below to support this claim. The compacter takes a stick diagram and produces compacted mask data from it. Compaction in HILL is constraint-based. We shall discuss how to extract a minimum system of constraints from the symbolic layout and how to solve constraint systems efficiently. The HILL simulator is a switch-level simulator. We discuss the underlying mathematical model of MOS circuit behaviour and show how to derive an efficient simulation algorithm from the model. The simulator is correct with respect to the model.

VLSI design systems are used to implement digital systems, i.e. to realize algorithms in hardware. Section 3 is devoted to efficient VLSI algorithms for the basic arithmetic functions. In particular, we will describe a multiplier for n bit binary numbers which has area $A = O(n^2)$, delay $T = O(\log n)$ and period $P = O(1)$.

Is this a good design? We can infer from Section 2 on VLSI complexity theory that it is. One of the results derived in that section and due to [BK 81, V 80] is that $AP^2 = \Omega(n^2)$ for every chip that can multiply. More generally, we shall present a theoretical model of VLSI computation and methods for deriving lower bounds on the complexity of concrete problems such as multiplication and addition. Moreover, we will compare the relative efficiency of various modes of computation, namely deterministic vs. randomized, in that model.

Section 5 contains a short survey of results obtained since the submission of this paper.

2 A COMPLEXITY THEORY FOR VLSI

This section is devoted to a complexity theory for VLSI computations. It is based on a theoretical model of VLSI computations which captures the essential features of the technology, in particular its planarity, but abstracts from the technological details. With respect to this model we derive two types of results:

(a) lower bounds on the complexity of important functions, e.g. shift and multiplication;
(b) relations between different models of computation, e.g. deterministic vs. randomized computation, influence of the I/O-convention, influence of the propagation delay assumption.

VLSI complexity theory originates with Thompson's Ph.D. thesis [T 80], which contains Theorems 1, 2 and 4 below. A number of researchers later extended his results. Specific references are given below.

2.1 The VLSI Model

Our VLSI model is based on Boolean circuits. This choice is adequate also for modelling more general "multidirectional" VLSI structures, e.g. buses [LM 81].

Definition 1. A *chip* $\chi = (\Gamma, \Lambda, \Delta)$ consists of the following structures.

(*a*) The *circuit* Γ: a synchronous Boolean circuit with feedback and unbounded fanin. Formally, this is a directed bipartite graph

$\Gamma = (V, E)$, where V is partitioned into a set S of *switches* and a set W of *wires*. Here $S = P \cup G$, where P is a set of *ports* labelled $\overline{\text{in}}$ or $\overline{\text{out}}$ and G is a set of *gates* labelled $\overline{\text{and}}$, $\overline{\text{or}}$, $\overline{\text{nand}}$, or $\overline{\text{nor}}$. For $s \in W$, if $(s, w) \in E$ then w is called an *output* of s, if $(w, s) \in E$ then w is called an *input* of s. All gates have out-degree 1, all wires have in-degree 1. Each input port has one input and one output. Each output port has two inputs and no output. The "additonal" input signal for the ports is an enable signal computed on the chip that activates the port.

(*b*) The *layout* Λ: the layout maps every vertex in the circuit into a compact connected region in the plane. Furthermore, each point in the region lies inside some Cartesian square of side length $\lambda > 0$ that is completely contained in the region. (This provision models the finite resolution of the fabrication process for VLSI chips.) Each point in the plane belongs to the interior of at most $\nu \geq 2$ regions. (The parameter ω is another fabrication-process-specific constant representing the number of functional *layers* on the chip. Since $\nu \geq 2$ non-planar circuits Γ can also be laid out.) Two regions touch exactly if the vertices they represent are neighbours in Γ. (We say that regions R_1 and R_2 touch if $R_1 \cap R_2 \neq \emptyset$ but $R_1^0 \cap R_2^0 = \emptyset$, where R_1^0, R_2^0 are the interiors of R_1, R_2.)

(*c*) The *manual* Δ: the manual is a set of directions for the communication between the chip and its environment. It contains for every input port a sequence of numbers that identify the input bits that enter through this port. The sequence also determines the order in which the input bits enter. When its enable signal is raised the port requests the next input bit in the sequence to enter. Analogously, for each output port the manual contains a sequence of numbers identifying the output bits produced at that port and their order. When its enable signal is raised the port produces the next output bit in the sequence.

After defining all components of the chip, we can define the operation of the circuit. To this end we associate with each port a word $w \in B^*$, $B = \{0, 1\}$, that we call its *history*. Furthermore, we label each wire with an initial Boolean value from $B \cup \{X\}$. (X stands for the undefined Boolean value, $0 \wedge X = 0$, $1 \vee X = 1$, $0 \vee X = 1 \wedge X = X$.) Such a labelling we call a *state* of the circuit. The initial state is most often the completely undefined state. In the ith cycle the circuit does the following. Each gate "reads" the values on all its input wires and computes the Boolean operation given by its label. The resulting value is put on its output wire. Each input port puts an X on its output wire if its enable signal is 0, otherwise it puts the next bit from its history on its output wire. Each output port checks if its enable signal is 1, and if so it puts its other input at the end of its history. Thus input ports consume their histories and output ports produce them. All actions of the gates happen in parallel. Thus a new state is reached on which the $(i + 1)$th cycle of the computation is started.

A VLSI computation uses up computational resources. We are interested in area A, time T and switching energy E. Area A is the area of the smallest rectangle that encloses layout Λ, and T is the number of steps taken by the circuit to produce the desired outputs. Alternative definitions of area and time and a definition of switching energy are discussed in Section 2.3.

We are now in a position to outline the argument for proving lower bounds on the AT^2 complexity of VLSI chips. Let us consider a chip computing some Boolean function $f: B^n \to B^m$. Let r be a smallest enclosing rectangle, let a, b be the side lengths of R, $a \leqq b$. Then $A = a \cdot b \geqq a^2$. Let us assume furthermore that the chip has n input ports and that a unique input bit is assigned to each port. Clearly, we can cut the chip into two halves L and R by a line c parallel to the side of length A of the chip, such that about half the input ports lie on either side of the cut. Then C has length $a \leqq A^{1/2}$ and hence at most $(2\nu/\lambda)A^{1/2}$ circuit components can intersect C. This can be seen as follows. Consider a strip of width 2λ with centre C:

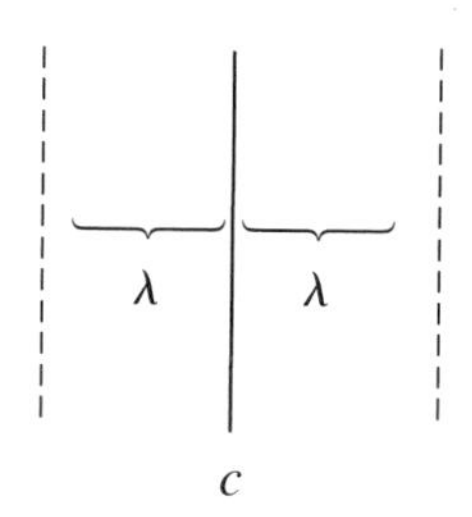

If a circuit component intersects C then it has a square of area at least λ^2 in common with the strip. Since any point of the strip belongs to at most ν (regions associated with) circuit components we conclude that $h\lambda^2 \leqq a2\lambda\nu$, where h is the number of circuit components intersecting c. Thus $h \leqq 2a\nu/\lambda$. We conclude further (and this is made precise in Theorem 1 below) that at most $h \leqq (2\nu/\lambda)A^{1/2}$ bits of information can cross c in any clock cycle.

Suppose now that we can show that ω bits of information have to cross cut C in order to allow successful computation of f. (We shall see below how such a claim can be shown.) Then the computation of f must take at least ω/h clock cycles, i.e. $T \geqq \omega/h \geqq \omega\lambda/2\nu A^{1/2}$ or $AT^2 \geqq (\lambda^2/4\nu^2)\omega^2$.

This concludes the basic lower-bound argument. We shall fill in the details in the next subsection.

2.2 Communication Complexity

The lower bound of the preceding subsection is based on the cost of communication in VLSI computations. We shall therefore study communication complexity of Boolean functions in somewhat more detail in this section.

Let $f: X \times Y \to A \times C$ be a function. We consider the following scenario. There are two computing agents L and R. Initially, L knows $x \in X$ and R knows $y \in Y$. They now want to cooperatively compute $f(x, y) = (a, c)$ by exchanging information between each other. More precisely, L sends a bit depending on x to R, R returns a bit depending on y and the bit just received, . . ., until L knows a and R knows c.

Definition [Y 79]. A deterministic algorithm is given by two response functions $r_L: X \times B^* \to B$ and $r_R: Y \times B^* \to B$ and two partial output functions $\text{out}_L: X \times B^* \to A$, $\text{out}_R: Y \times B^* \to C$, where $B = \{0, 1\}$. A computation on input x, y is a sequence $w = w_1 w_2 \ldots w_k$ of bits such that

(1) $w_{2i+1} = r_L(x, w_1 \ldots w_{2i})$ for $i \geqq 0$,
(2) $w_{2i+2} = r_R(y, w_1 \ldots w_{2i+1})$ for $i \geqq 0$,
(3) $(x, w_1 \ldots w_k) \in \text{dom}(\text{out}_L)$ and
$(y, w_1 \ldots w_k) \in \text{dom}(\text{out}_R)$,
(4) there is no shorter sequence with this property.

k is the length of the computation w and is denoted $k(x, y)$. An algorithm is correct if $f(x, y) = (\text{out}_L(x, w), \text{out}_R(x, w))$, where w is the computation on input x, y, for all $x \in X$, $y \in Y$. The complexity of an algorithm Alg is defined as

$$C(\text{Alg}) = \max \{k(x, y); x \in X, y \in Y\}.$$

Finally, the complexity of f is defined by

$$C_{\text{def}}(f) = \min \{C(\text{Alg}); \text{Alg computes } f\}.$$

For the definitions above we have assumed that the partition of the inputs and outputs into left and right inputs and outputs is part of the problem specification. This assumption is quite reasonable in applications to distributed computing in general; it is, however, too restrictive for VLSI computations. Note that in VLSI a problem is given as a Boolean function $f: B^n \to B^m$. It is up to the chip designer to fix the locations on the chip where certain inputs are consumed and certain outputs produced. We therefore make the following definition.

Definition. Let $f: B^n \to B^m$ and let $\omega \geqq 0$. Function f is ω-separable if for all balanced partitions X, Y of $[1 \ldots n]$, i.e. $n/3 \leqq |X|, |Y| \leqq 2n/3$, and all partitions A, C of $[1 \ldots m]$ we have $C_{\mathrm{def}}(f_{X,Y,A,C}) \geqq \omega$, where $f_{X,Y,A,C}$: $B^X \times B^Y \to B^A \times B^C$ is defined by partitioning the input and output bits into left and right input and output bits as given by partitions X, Y and A, C.

We can now formalize the first part of the lower-bound argument. In the form given here, Theorem 1 emerged over a sequence of papers [T 80, BK 81, V 80, S 81, LS 81, K 82].

Theorem 1. If $f: B^n \to B^m$ is ω-separable then $AT^2 \geqq (R^2/16r^2)\omega^2$ for every chip computing f.

Proof. Let $\chi = (\Gamma, \Lambda, \Delta)$ be a chip computing f. Let π be an input port and let R_π be its associated region in the layout Λ. Let the input bit x_i enter the chip through input port π. With each such x_i we associate a point p_i in the interior of R_π such that different points are associated with different input bits. For the purposes of the lower-bound proof we will consider the bit x_i to enter the chip through point p_i. Similarly, we associate a point q_i with every output bit y_i.

Let Q be a smallest-area rectangle enclosing layout Λ. Then Q has side lengths a, b. Assume $a \leqq b$. We can cut Q into halves L and R by a cut C parallel to the side of length a such that exactly half of the p_is lie to the left of cut C. Cut C gives rise to a balanced partition X, Y of the input bits and a partition A, C of the output bits in a natural way. Let $f' = f_{X,Y,A,C}$ be the function induced by these partitions. Since f is ω-separable we conclude that $C_{\mathrm{det}}(f') \geqq \omega$.

Since cut C has length at most a, at most $h \leqq (2r/\lambda)A^{1/2}$ regions of Λ associated with components of Γ can intersect C. We shall now derive from the chip an algorithm Alg for f' with $C(\mathrm{Alg}) \leqq 2Th$.

Consider computation of f by χ. At each cycle we associate two values with C, a left and a right "crossing" value. The left (right) crossing value $v^\ell = (v_1^\ell, \ldots, v_h^\ell)$ $(v^r = (v_1^r, \ldots, v_h^r)$ contains a component $v_i^\ell(v_i^r)$ for each region R_i intersecting C. If R_i is a $(\overline{\mathrm{nand}}, \overline{\mathrm{nor}}, \overline{\mathrm{and}}, \overline{\mathrm{or}})$ gate then $v_i^\ell(v_i^r)$ is the (nand, nor, and, or) of all input values during the last cycle whose regions intersect $L(R)$. If R_i is a wire then $v_i^\ell(v_i^r)$ is the above value for its input gate. Ports act as *and*-gates in this context.

With these definitions the computation of f by χ can be regarded as a deterministic algorithm for computing f' in the sense of the definition above. The information exchanged between L and R are the crossing values. The left crossing values are sent from L to r, and the right crossing values are sent from R to L. The computation is completed when both sides have produced their outputs. Since a total of $2k$ bits are exchanged in every cycle the algorithm described above has complexity $2Th$.

Since $2Th \geqq C(\mathrm{Alg}) \geqq C_{\det}(\delta') \geqq \omega$ we conclude that

$$AT^2 \geqq (\lambda^2/16v^2)\omega^2. \qquad \square$$

We shall next derive methods for proving lower bounds on the communication complexity of functions. We shall discuss the crossing sequence method for multiple-output functions and the rank method for single-output functions.

Method 1 (*crossing sequences for multiple-output functions*)

Definition. Let $f: X \times Y \to B^m$ and let I, J be a partition of the output bits of f. f has ω-flow if there is partial input $y \in Y$ such that f restricted to $X \times \{y\}$ in its domain and J in its range has more than $2^{\omega-1}$ different points in its range.

Theorem 2. If f has ω-flow then $C_{\det}(f) \geqq \omega$.

Proof. Assume that there is a deterministic algorithm computing f that has a communication length of less than ω. Then for two inputs (x_1, y), (x_2, y) generating different output configurations in J the same communication sequence w is generated. Thus for some $j \in J$, $f_j(x_1, y) \neq f_j(x_2, y)$, but the algorithm computes $f_j(x_1, y) = \mathrm{out}_{R,j}(y, w) = f_j(x_2, y)$. Here f_j is the jth bit of function f, and similarly for output function f_R. Thus the algorithm does not compute f correctly, a contradiction. $\square$

[V 80] gives an example of a class of functions to which Method 1 applies.

Definition 2. Let $f(x_1, \ldots, x_n, s_1, \ldots, s_m) = (y_1, \ldots, y_n)$ be a Boolean function. f computes a permutation group G on n elements if for all g $\in G$ there is an assignment $\alpha_1, \ldots, \alpha_m$ to the $s_1, \ldots, s_m$ such that $f(x_1, \ldots, x_n, \alpha_1, \ldots, \alpha_m) = (x_{g(1)}, \ldots, x_{g(n)})$ for all choices of $x_1, \ldots, x_n$. We call $x_1, \ldots, x_n$ the permutation inputs and $s_1, \ldots, s_k$ the control inputs. f is called *transitive* of degree n if G is a transitive group, i.e. if for all $i, j = 1, 1, \ldots, n$ there is a $g \in G$ such that $g(i) = j$.

The most straightforward example of a transitive function of degree n is the cyclic shift function cs $(x_1, \ldots, x_n, s_1, \ldots, s_m) = (y_1, \ldots, y_n)$, where $n = 2^m$ and the $s_1, \ldots, s_m$ encode a number k, $0 \leqq k < n$, and $y_1 = x_{(i+k) \bmod n}$. cs computes the transitive group of cyclic permutations. Other examples of transitive functions of degree $\Omega(n)$ are the multiplication of n-bit integers, the multiplication of three $n^{1/2} \times n^{1/2}$ matrices, and the sorting of n numbers between 0 and n.

Theorem 3. Let $f: B^{n+m} \to B^n$ be transitive of degree n. Then f is $n/6$-separable.

Proof. Let G be the transitive group computed by f. The equivalence relation $g(i) = h(i)$ for fixed but arbitrary $i \in \{1, \ldots, n\}$ divides G into n equivalence classes of size $|G|/n$. Let A be the set of all permutation input bits and let B be the set of all output bits. Let X, Y be any partition of $A \cup B$ such that $|X|, |Y| \leqq 4n/3$. For each input bit i in X and output bit j in Y there are $|G|/n$ group elements $g \in G$ such that $g(i) = j$. Let w.l.o.g. X be no greater than Y assume that X contains at least as many input bits as output bits. (The other cases can be argued similarly.) Let S be the set of input bits in X and let S' be the set of ouput bits in Y. Then

$$|S| \cdot |S'| \geqq (n/3) \cdot (n/2) = n^2/6.$$

For each of the pairs $(i, j) \in S \times S'$ there are G/n group elements matching them. Since there are only a total of $|G|$ group elements there must be one element $g_0 \in G$ realizing at least $n/6$ matchings between inputs in S and outputs in S'. The partial input y realizing the flow sets the control input bits in Y such that together with appropriate assignments to the control input bits in X they encode this element g_0. The other input bits in Y are assigned arbitrarily. □

Theorem 4 [T 80, BK 81]. There is a constant $c > 0$ such that for every chip computing the cyclic shift of n inputs or multiplying n-bit binary numbers $AT^2 \geqq cn^2$.

Proof. It was shown above that the cyclic shift function is transitive of degree n. Thus the claim follows from Theorems 1 and 2. In order to extend the result to multiplication, we only have to notice that multiplication by a power of two is a shift. □

Theorem 4 is quite significant because it states a lower bound on AT^2 for two very important functions: cyclic shift and multiplication. The $AT^2 = \Omega(n^2)$ lower bound is a yardstick against which one can measure actual design. This will be done in Section 3.

We shall now turn to the rank method for proving lower bounds on the communication complexity of Boolean predicates.

Method 2 (the rank lower bound for Boolean predicates)

Let $p: X \times Y \to B$ be a predicate. With $A = B$ and $C = \{1\}$ we can use the definitions above to define $C_{\det}(p)$. There are two methods for proving lower bounds on $C_{\det}(p)$: the crossing sequence method and the rank method. The former is older, easier to apply, and similar to the method used to prove Theorem 2. Since the rank method is more general we shall describe it here. The rank method was developed in [MM 82].

Definition. Let r be a ring and let $r^{(n,m)}$ be the set of $n \times m$ matrices over r. The *rank* of $A \in r^{(n,m)}$ over r is the minimum k such that A can be written as $A = C \cdot D$, where $C \in r^{(n,k)}$ and $D \in r^{(k,m)}$. We use N to denote the ring of integers.

If r is a field the above definition coincides with the definition of matrix rank known from linear algebra. Method 2 is based on the following theorem.

Theorem 5. Let $p: X \times Y \to B$ be a Boolean predicate, and let P be its associated matrix, i.e. P is an $|X|$ by $|Y|$ matrix with $P_{x,y} = p(x, y)$. Then

$$C_{\det}(p) \geqq \log \operatorname{rank}_N (P) \geqq \log \operatorname{rank}_r (P),$$

where r is any field.

Proof. The second inequality is known from algebra. For the proof of the first inequality we state the following lemma.

Lemma 1. let r be a ring, $A \in r^{(n,m)}$, $B \in r^{(n,k)}$, $C \in r^{(k,m)}$. Then

$$\operatorname{rank}_r ((A \quad B)) \leqq rank_r (A) + \operatorname{rank}_r (B),$$

$$rank_r \left(\begin{pmatrix} A \\ C \end{pmatrix} \right) \leqq rank_r (A) + \operatorname{rank}_r (C).$$

Proof. If $A = D_1 \cdot E_1$ and $B = D_2 \cdot E_2$ then

$$(A \quad B) = (D_1 \quad D_2) \cdot \begin{pmatrix} E_1 & 0 \\ 0 & E_2 \end{pmatrix}.$$

The proof of the second inequality is analogous. □

Now consider any deterministic algorithm for computing p. Inductively on the length of $w \in B^*$ we define the matrix p_w as follows. For $|w| = 0$,

$P_\varepsilon := P$. For $|w| > 0$, if $|w| = 2l$ then P_{w0} (P_{w1}) is obtained from P_w by selecting all rows x with $r_L(x, w) = 0$ ($r_L(x, w) = 1$). If $|w| = 2l + 1$ then P_{w0} (P_{w1}) is obtained from P_w by selecting all columns y with $r_R(y, w) = 0$ ($r_R(y, w) = 1$).

By Lemma 1 we have $\max(\text{rank}_N(P_{w0}), \text{rank}_N(P_{w1})) \geqq \text{rank}_N(P_w)/2$. Moreover, if $\text{out}_R(x, w)$ is defined then rank $P_w \leqq 1$, since P_w must consist of a set of rows that are constant 0 and a set of rows that are constant 1. Thus there are $x \in X$, $y \in Y$ such that the computation of x, y has length at least $\log \text{rank}_N(P)$. □

We shall next give two applications of Theorem 5.

Theorem 6. Let $X = Y$ and let $p(x, y) = (x = y)$ be the identity predicate. Then $C_{\text{det}}(p) \geqq \log |X|$.

Proof. Clearly, P is the identity matrix and hence $\text{rank}_N(P) \geqq |X|$. □

The second example is less trivial and illustrates the fact that randomization helps in distributed computing and in VLSI. In *Las Vegas* computations computing agents L and R have fair coins available to them. The response of an agent, say L, depends on his argument, on the history of the computation and on the outcome of a toss of the coin. Correctness of an algorithm is defined as above, i.e. the output of the computation must be independent of the outcomes of the coin tosses. The complexity of an algorithm on input x, y is the expected number of bits exchanged. A precise definition is as follows.

Definition. A Las Vegas algorithm is given by two response functions $p_L: X \times B^* \times B^t \to B$ and $p_R: Y \times B^* \times B^t \to B$ and the partial output function $a: B^* \to B$. We assume that both L and R first toss t coins to determine the third arguments t_L, t_R in the response functions, and then start a deterministic computation. The computation ends when $(w_1, \ldots, w_{k(x,y,t_L,t_R)}) \in \mathit{dom}(a)$. Its result is $a(w_1, \ldots, w_k)$.

The Las Vegas communication complexity of p is

$$C_{\text{LV}}(p, L \leftrightarrow R) = \min_{\substack{A \\ \text{LV-alg}}} \sum_{t_L, t_R \in B^t} k(x, y, t_L, t_R)/2^{2t}.$$

We consider the following example.

Definition. Let $n \in N$ and $X = Y = [0:2^n - 1]^n$. For $x = (x_1, \ldots, x_n) \in X$ and $y = (y_1, \ldots, y_n) \in Y$, let

$$p_1(x, y) = \begin{cases} 1 & \text{if } x_i = y_i \text{ for some } i, 1 \leqq i \leqq n, \\ 0 & \text{otherwise.} \end{cases}$$

Theorem 7. (*a*) $C_{\text{det}}(p_1) \geqq n^2$,
(*b*) $C_{\text{LV}}(p_1) = 0(n(\log n))$.

Proof. (*a*) Since $P_1 \in B^{2^{(n^2)}}$ we only have to show that $\text{rank}_{\text{GF}(2)} P_1 \geqq 2^{(n^2)}$, where GF(2) is the field of characteristic 2. Let $\oplus$ denote addition modulo 2. We transform the matrix $\overline{P}_1$ associated with $\overline{p}_1$ into the identity matrix of size $2^{(n^2)} \times 2^{(n^2)}$ by means of linear transformations.

Lemma 2. Let $w_1, \ldots, w_n, y_1, \ldots, y_n \in [0:2^n - 1]$. Define

$$g(w_1, \ldots, w_n, y_1, \ldots, y_n) := \sum_{\substack{x_1 \\ x_1 \neq w_1}} \cdots \sum_{\substack{x_n \\ x_n \neq w_n}} \overline{p}_1(x_1, \ldots, x_n, y_1, \ldots, y_n).$$

Then $g(w_1, \ldots, w_n, y_1, \ldots, y_n) = \text{id}\,(w_1, \ldots, w_n; y_1, \ldots, y_n)$.

Proof. Note that

$$\begin{aligned} g(w_1, \ldots, w_n, y_1, \ldots, y_n) &= |\{(x_1, \ldots, x_n); x_i \neq w_i, x_i \neq y_i\}| \bmod 2 \\ &= \prod_{i=1}^{n} (2^n - |\{y_i, w_i\}|) \bmod 2 \\ &= 1 \end{aligned}$$

iff $y_i = w_i$ for all i.

We conclude from Lemma 2 that $\text{rank}_{\text{GF}(2)} \overline{P}_1 = 2^{(n^2)}$ and hence $C_{\text{det}}(p_1) = C_{\text{det}}(\overline{p}_1) \geqq n^2$.

(*b*) The Las Vegas algorithm for p is based on the following simple number-theoretic fact.

> Let $p_1, p_2, \ldots, p_m$ be the set of primes $\leqq n$.
> Let $0 \leqq x, y \leqq 2^n - 1$. If $x \neq y$ then
> $|\{i; x \bmod p_i \neq y \bmod p_i\}| \geqq m/2$.

The algorithm is as follows:

```
for i from 1 to n
do for k from 1 to log n
   do L selects a prime p_j from the list of
      primes p_1, ..., p_m ≦ n at random and sends
      (p_j, x_i mod p_i) to R;
      r computes y_i mod p_j;
      if x_i mod p_j ≠ y_i mod p_j
      then goto nexti fi
```

```
        od
        L sends x_i to R;
        if x_i = y_i then halt and output 1;
        next i :
    od
    halt and output 0;
```

The algorithm above is clearly correct. Also note that if $x_i = y_i$ then L will send $O((\log n)^2 + n)$ bits to r until this fact is detected. Observe that case $x_i = y_i$ occurs almost once. If $x_i \neq y_i$ then $O(k \log n)$ bits are sent from L to R with probability 2^{-k} and an additional n bits are sent with probability $2^{-\log n} = 1/n$. Thus an expected number $O(\Sigma_{k \geqq 1} k \cdot 2^{-k} \log n + (1/n) \cdot n) = O(\log n)$ bits are sent. Hence $C_{\mathrm{LV}}(p_1) = O(n \log n)$. □

Theorem 7 is quite significant. For predicate p_1 randomization *provably* reduces the amount of communication required by almost a square root. How about chip complexity? Note first that it is conceivable to incorporate random devices into VLSI chips. Such a device might use statistical-physical effects to produce (true?) random sequences. Let us assume that we can build a device that uses area $O(1)$ and produces a random bit in time $O(1)$. A predicate similar to p_1 can be used to show the following (cf. [MM 82] for details).

Theorem 8. There is a predicate $p : B^n \to B$ such that $AT^2 \geqq cn^2$ and $(AT^2)_{\text{Las Vegas}} \leqq cn^{3/2}(\log n)^3$ for some constant c.

2.3 Extensions and Related Results

In this subsection we shall briefly mention some extensions and some related results.

2.3.1 Area

We defined the area A of a chip as the area of the smallest enclosing rectangle. Alternatively and more naturally we might define A as the area of the union of the regions associated with circuit components. Let us call this area the active area. In [LM 81] it is shown that all AT^2 lower bounds are valid with area replaced by active area.

2.3.2 The manual and lower bounds on area

The manual is a set of directions for the communciation between the chip and its environment. Manuals as defined above were termed strongly where-oblivious manuals in [LM 81]. A more restricted class of manuals are the when- and where-oblivious manuals. In these manuals the location and the time at which a bit enters the chip is independent of the input. We have the following theorem.

Theorem 9. Let f be a transitive function of degree n. Then

(*a*) [V 80] a chip for f has area $A = \Omega(n)$ if the manual is where- and when-oblivious;

(*b*) [LM 81] a chip for f has area $A = \Omega(n^{1/3})$ if the manual is strongly where-oblivious.

2.3.3 *Period*

The period P of a chip is the least distance in time between distinct problem instances that can be fed into the chip. Vuillemin [V 80] has shown that the $AT^2 \geqq n^2$ lower bound of Theorem 4 can be strengthened to $AP^2 = \Omega(n^2)$. Baudet [Ba 81] has shown that $AP = \Omega(n + (n \log n)/T)$ for every chip realizing binary addition.

2.3.4 *Energy*

Switching energy E is another important computational resource. We assume that every unit of active chip area consumes one unit of switching energy each time it changes its state from 0 to 1 or vice versa.

This complexity measure is closely related to the energy dissipated when charging a wire. In technologies without high d.c. currents the switching energy dominates the total energy dissipation on the chip. We have the following theorem.

Theorem 10 [LM 81]. Let f be a transitive function of degree n. Let E be the worst-case switching energy consumed by any chip computing f. Let A be the (active) chip area and let T be the worst case computing time. Then

$$c_1 AT^2 \geqq ET \geqq \frac{c_2 n^2}{\log \dfrac{c_3 AT^2}{n^2}} \geqq 0,$$

for appropriate constants c_1, c_2, $c_3 > 0$.

2.3.5 *Propagation delay and the notion of time*

We defined time t as the number of clock cycles spent on the computation. Note that this is not a "physical" measure of time, since the length of a clock cycle may itself vary with the size of the chip. However, as long as the delay of signal propagation along wires is not significantly longer than the delay of the switching elements, the number of clock cycles gives a good representation of the time spent, i.e. is asymptotically accurate. As the size of chips increases this ceases to be the case, at least if the driving capacity of transistors driving long wires is not increased appropriately. [CM 81] introduce a physical time measure T by measuring time in seconds under the assumption that signals are propagated along wires at a constant speed. They get dramatically different lower bounds on circuit complexity. Not only their lower bounds are larger, as we expect, but the optimal chips according to their complexity measures differ significantly from the optimal chips if T is measured in clock cycles. This is because [CM 81] pay a penalty for long wires across which communication is expensive. The time measure of [CM 81] may become technologically significant eventually, as the speed of light becomes the limiting factor in signal propagation on VLSI chips. In the meantime, however, other physical time measures may be more appropriate, such as the one introduced in [MC 80] that is based on the capacitive properties of VLSI structures. In this model the switching time of a transistor is given by the ratio of the capacity to be driven and the size of the driving transistor. No non-trivial lower-bound results have been shown for this model as of today. However, in many cases optimal designs for the unit delay model carry over to the capacity model.

3 EFFICIENT VLSI ALGORITHMS

In Section 2 we derived lower bounds on the complexity of the basic arithmetic functions. In particular,

$$AP^2 = \Omega(n^2), \qquad T \geqq P \text{ for binary multiplication,}$$
$$AP = \Omega(n + (n \log n)/T) \qquad T \geqq P \text{ for binary addition.}$$

In both cases there are designs whose performance matches the lower bound.

Theorem 1. (*a*) ([PV 81]) For $n^{1/2} \geqq T \geqq (\log n)^2$ there is a chip for binary multiplication with $AT^2 = O(n^2)$, $P = T$.
(*b*) ([BK 81]) For $n \geqq T \geqq \log n$ there is a chip for binary addition with $AP = O(n + (n \log n)/T)$.

We will not go into these constructions here. Rather we shall describe a fast $T = O(\log n)$ and area efficient multiplier. More precisely, we describe
(*a*) a chip with $A = O(n^2)$, $T = O(\log n)$, $P = O(1)$, i.e. $AP^2 = O(n^2)$;
(*b*) a chip with $P = T = O(\log n)$ and $A = O(n^2/\log n)$.
Previously, only designs with $A = O(n^2 \log n)$, $T = O(\log n)$ and $P = O(1)$ were known [V 83, Be 82].

Our design is based on the following well-known identity due to Karatsuba. Let $a = a_1 2^{n/2} + a_2, b = b_1 2^{n/2} + b_2$ be two n-bit integers; a_1, a_2, b_1, b_2 are $n/2$-bit integers. Let $h = (a_1 + a_2)(b_1 + b_2)$. Then we have $c = a \cdot b = a_1 b_1 2^n + (h - a_1 b_1 - a_2 b_2) 2^{n/2} + a_2 b_2$. Thus multiplication of 2 n-bit numbers can be reduced to three multiplications of $n/2$-bit numbers and a few additions.

Let $T(n)$ be the delay of a network based on the identity described above. We have

$$T(n) = T(n/2) + O(A(n)),$$

where $A(n)$ is the delay of the adder used. If one uses a fast adder, e.g. a carry lookahead adder, then $A(n) = O(\log n)$ and hence $T(n) = O((\log n)^2)$. A design based on these principles can be found in [Lu 81].

We add one more idea to Luk's design: a redundant number of representation. Then $A(n) = O(1)$ and hence $T(n) = O(\log n)$ if the representation is chosen appropriately. Redundant representations were also used in [Be 82, V 83]. For $a_0, \ldots, a_{n-1} \in \{-1, 0, 1\}$ let

$$\text{num}\ (a_{n-1}, \ldots, a_0) = \sum_{i=0}^{n-i} a_i 2^i;$$

i.e. we represent numbers with digits $-1, 0, +1$. We shall next show how to add two numbers in this representation in time $O(1)$.

Let $a_0, \ldots, a_{n-1}, b_0, \ldots, b_{n-1} \in \{-1, 0, 1\}$ and let $a = \text{num}\ (a_{n-1}, \ldots, a_0)$, $b = \text{num}\ (b_{n-1}, \ldots, b_0)$ and $c = a + b$. Note first that $a_i + b_i \in \{-2, -1, 0, +1, +2\}$. Next write $a_i + b_i$ as $2n_i + s_i$, where n_i and s_i are given in Table 1. Finally, define $c_i = n_{i-1} + s_i, 0 \leqq i \leqq n$.

Lemma (*a*) $c_i \in \{-1, 0, +1\}$
(*b*) $c = a + b = \text{num}\ (c_n, c_{n-1}, \ldots, c_0)$.

Table 1.

$a_i + b_i$	n_i	s_i	
2	1	0	
1	0	1	if $a_{i-1} + b_{i-1} < 0$
	1	-1	if $a_{i-1} - a_{i-1} \geqq 0$
0	0	0	
-1	0	-1	if $a_{i-1} + b_{i-1} > 0$
	-1	$+1$	if $a_{i-1} + b_{i-1} \leqq 0$
-2	-1	0	

Proof. (*a*) Note first that n_{i-1}, $s_i \in \{-1, 0, 1\}$. If $a_i + b_i \neq \pm 1$ then $s_i = 0$ and hence $c_i \in \{-1, 0, 1\}$. So let us assume that $a_i + b_i = +1$, the case $a_i + b_i = -1$ being symmetric. If $a_{i-1} + b_{i-1} < 0$ then $u_{i-1} \leqq 0$ and $s_i = 1$. Hence $c_i \in \{0, 1\}$. If $a_{i-1} + b_{i-1} \geqq 0$ then $u_{i-1} \in \{0, 1\}$ and $s_i = -1$. Hence $c_i \in \{-1, 0\}$. In either case we have shown that $-1 \leqq c_i \leqq 1$.

(*b*) is obvious. □

We represent digits in $\{-1, 0, +1\}$ by two bits. More precisely, we use representation in 1-complement, i.e. encoding $(+1) = (0, 1)$, encoding $(0) = (0, 0)$ encoding $(-1) = (1, 0)$. Then it is easy to design an adder cell for n-digit numbers with height $O(1)$ and width $O(n)$

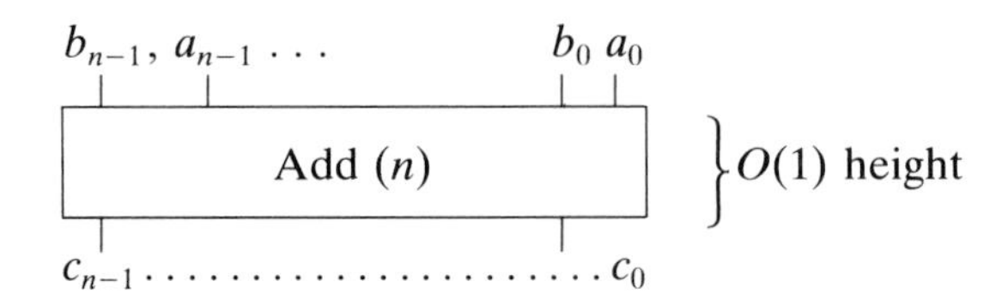

Using this adder cell, we can build up a multiplier as shown in Figure 1. A conventional multiplier is used for small n, say $n \leqq 8$. Let $H(n)$ ($W(n)$) be the vertical (horizontal) extension of that layout. Then

$$H(n) = O(n) + 3W(\lceil n/2 \rceil + 1),$$
$$W(n) = O(n) + H(\lceil n/2 \rceil + 1),$$

also $H(n) = W(n) = cn$ for $n \leqq 8$ and some appropriate constant c.

We conclude that $H(n) = O(n) + 3H(\lceil(\lceil n/2 \rceil + 1)/2\rceil + 1)$, which has solution $H(n = O(n)$. Thus $W(n) = O(n)$, and hence $A(n) = H(n) \cdot W(n) = O(n^2)$.

Theorem 2. The multiplication network described above has area $A = O(n^2)$, delay $T = O(\log n)$ and period $P = O(1)$. In particular, $AP^2 = O(n^2)$, which is optimal.

Proof. The circuit described above computes the product of two n-bit numbers in redundant representation. It is easy to see how to use a $T = O(\log n)$ adder to convert from redundant representation to standard binary representation. This proves that $T = O(\log n)$ and $A = O(n^2)$. Finally, observe that the circuit above is synchronous and hence can be used in a pipelined fashion. Thus $P = O(1)$. □

We finally describe how to reduce the area. Let a, b be two n-bit numbers. Let $k = (\log n)^{-1/2}$. Divide a, b into $t = n/k$ pieces of length k each. Then

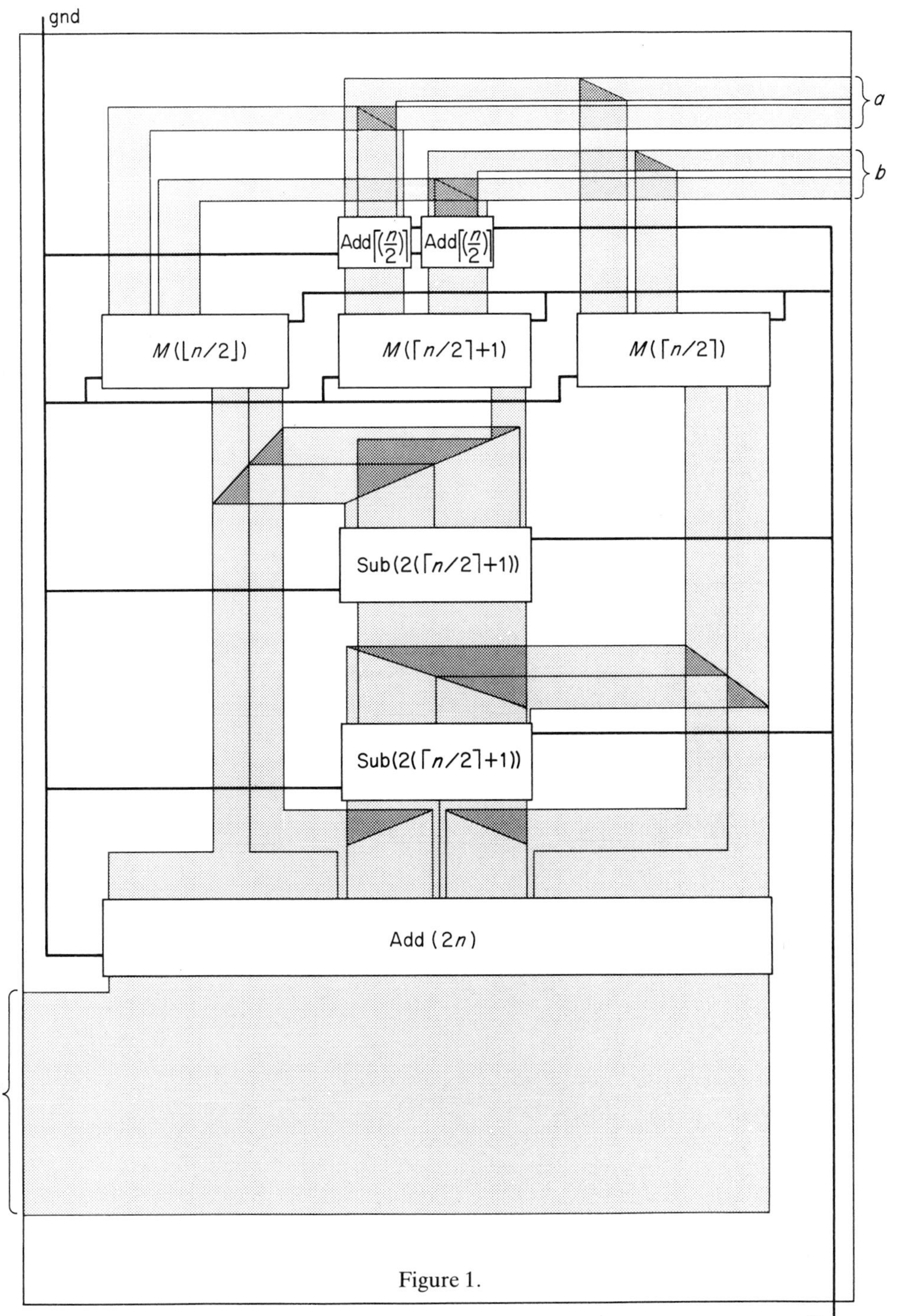

Figure 1.

$$a = \sum_{i=0}^{t} a_i 2^{ki}, \quad b = \sum_{i=0}^{t} b_i \, 2^{ki}$$

and each a_i, b_i is a k-bit binary number. Also

$$c = a \cdot b = \sum_{i=0}^{t} \sum_{j=0}^{t} a_i b_j 2^{(i+j)k}.$$

We use a k-bit multiplier to compute the t^2 products $a_i b_j$. Since the multiplier can be pipelined this takes time $O(\log n + t^2) = O(\log n)$. Also the t^2 products can clearly be added in time $O(t^2) = O(\log n)$ if redundant number representation is used. We summarize in the following theorem.

Theorem 3. There is an $A = O(n^2/\log n)$, $P = T = O(\log n)$ multiplier for n-bit binary numbers.

4 THE HILL DESIGN SYSTEM

The HILL design system is currently under development at the University of Saarbrücken. It is part of a larger VLSI project which is sponsored by the DFG (Deutsche Forschungsgemeinschaft). A major other project is the CADIC system, which is being developed under the direction of G. Hotz.

An overview over the HILL system is given in Figure 2. At present the HILL system has three major parts:

(*a*) HILL layout specification language and graphics editor (discussed in Subsection 4.1);
(*b*) HILL Compacter (Subsection 4.2);
(*c*) HILL Simulator (Subsection 4.3).

4.1 HILL Layout Language and Graphics Editor

HILL is a tool for single-chip development. The main focus of HILL is layout generation and verification. HILL aims at supporting the designer who has a comprehensive global image of his circuit. HILL provides a convenient way of describing a layout symbolically either by a HILL program or during an interactive session at the graphics terminal. Even though the layout is specified in a symbolic manner, the designer has many means for exerting direct influence on the quality of the resulting mask data.

HILL is a system that combines convenient circuit description with efficiency of the implementation. We aim for efficiency in three respects: human design time, chip area and delay, and computational resources.

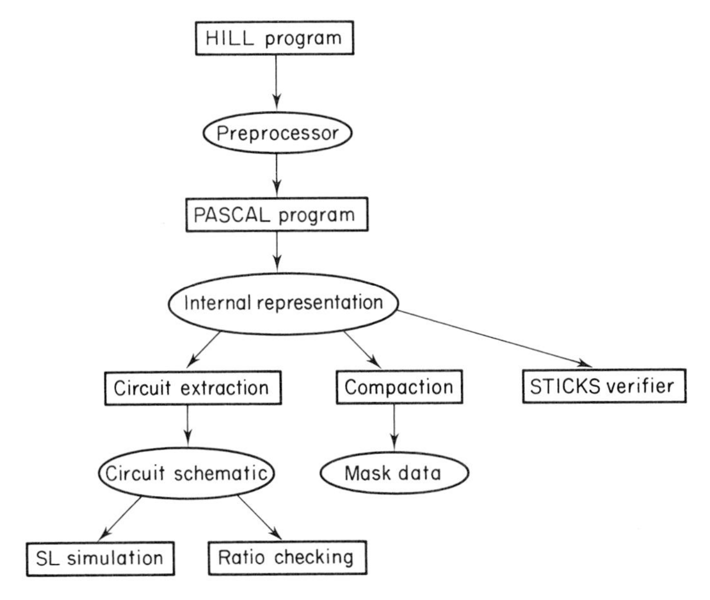

Figure 2 Overview of the HILL System.

Human resources. In HILL, integrated circuits are described at the level of stick diagrams enhanced by extensive means for structuring a design hierarchically. We have chosen the level of stick diagrams because on the one hand it still allows the designer to express his insights about the topology of the circuit and on the other hand it frees the designer from the tedious and error-prone task of specifying his circuit on the mask level. The stick diagram level has been used successfully in systems like [CABBAGE], [STICKS] and [MULGA].

Even though it certainly is no good practise to specify a whole large-scale circuit with one great stick diagram, if enhanced by extensive means for hierarchical structuring, especially with a powerful cell concept, stick diagrams become a convenient symbolic representation of even large-scale layouts.

In the HILL report [HILL 86] this thesis is exemplified by a number of examples. We shall give only a small example below.

Hierarchical structuring exploits the regularity of a layout. Of course, there are always irregular parts of a layout, and therefore HILL also allows graphical layout specification. Previous stick-diagram systems used to be totally or almost totally graphics-oriented. However, graphics alone cannot support powerful mechanisms for structuring layouts. Essentially, it can only support composition and simple forms of iteration. However, it cannot support recursion, and the full power of iteration and parameter passing.

Recursion and iteration are central to (software) algorithm design, and they have already proved to be powerful concepts in hardware design [GV 82, MC 80]. Only a high-level programming language provides the flexibility needed, and so HILL is designed as a PASCAL extension, which interacts gracefully with a graphics editor.

A good example is provided by the multiplier designed in the previous section. The layout given there is regular, however, the regularity can certainly not be captured in a pure graphics system. Rather, powerful descriptive tools, such as recursion, iteration and parameterized cells are needed to capture the regularity.

Chip area and delay. Experience with existing systems [CABBAGE, STICKS, MULGA, HILL 84] suggests that automatic compaction can yield small layouts which come very close to hand-compacted layouts. The stick diagram level is close enough to the silicon to allow the designer to incorporate performance aspects into his specification, and the compacter supports chip performance with his knowledge of the fabrication process. Finally, the case of circuit description and the flexibility of the algorithms used in the system allow the designer to try several approaches to his circuit and select the one he likes best.

Computational resources. Most existing systems have definite shortcomings in this respect. In all cases no theoretical analysis of the running time is given; often algorithmic concepts enter the system only scantily. Computational experience with the systems suggests that the running time is highly non-linear. For example, it is reported that CABBAGE takes time $O(n^{1.2})$ to compact a circuit with n transistors and wires. In the HILL prototype we improved upon this; an $O(n \log n)$ compaction algorithm is described in [Le 83]. However, even such an algorithm will not do for large-scale circuits, mainly because it also has an $O(n^{1/2})$ space requirement. The solution to this problem is to compact hierarchically; see [Le 82b]. The compaction algorithms used in HILL are described in Sub-section 4.2.

The main structuring device in HILL is that of a "*cell*". A cell is the specification of a subcircuit of the chip to be designed. This subcircuit will in general function as a module in the chip that communicates with its surroundings through relatively few connections (pins) and performs a specific subfunction of the chip function. It is rectangular in shape, with the connection pins arranged on its boundary. It is very much reminiscent of a function in a sequential programming language like PASCAL. As procedure parameters form the (up to side effects exclusive) interface between the procedure and its call environment, the pins of a cell facilitate the interface between the cell and the circuitry around its location of "placement" on the chip. The *only* way to contact to a cell is through one of its pins. Like procedures cells can be compiled separately and defined externally. For "instantiating" a cell only a description of its rectangular boundary, its so-called *template* has to be given. The template contains no (electrical or topological) information about any of the inner workings of the cell. However, it contains both electrical and topological information of the cell boundary. Pins can be related to each other electrically in the template, and they have to be "placed" in order to specify the order in which they occur around the cell boundary.

In addition to specifying in which order the pins appear on each side of the cell, HILL also requires to specify the relative positions of pins on opposite sides of a cell. This is in marked contrast with other symbolic layout systems, e.g. ALI, and has the following advantage. A layout (even a symbolic one) restricts the relative positions of pins; e.g. in the symbolic layout for the multiplier given in the previous section the output pins must be to the right of the input pins. In HILL these restrictions are part of the template, and hence can be taken into account by a designer when he uses a cell.

In the rest of this subsection we give an example that highlights many of the features of the HILL layout language. For a complete description the reader is referred to [HILL 86].

We specify the logical part of a decoder. The decoder takes n inputs (horizontally) and 2^n inputs at the top. It produces 2^n outputs at the bottom, all but one of which are zero. One of the inputs, namely the one selected by the horizontal inputs, is fed through.

For the design we need two leaf cells zero and one. The layouts of these cells are given below:

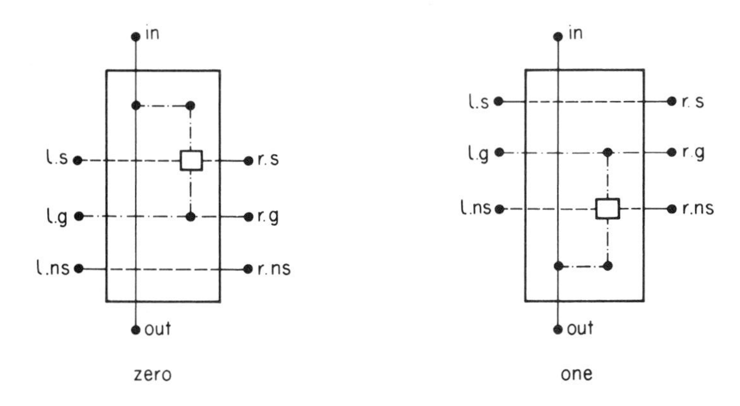

(The horizontal diffusion line carries gnd; lines s and ns carry a signal and its complement. Cell zero lets the signal fed in at the top pass through iff the signal is zero. Otherwise it grounds the signal.) Cells zero and one are most naturally defined graphically and then entered into the cell library. They can then be included into a HILL program by

```
aggregate bus = record s: poly; g: diff sig = gnd;
                                  ns: poly end;
cell zero = temp pins in, out: metal;
                      l,r: bus;
                order implicit all;
                sides top in;
                      bottom out;
                      left l;
                      right r;
            pmet;
            external;
```

In these definitions we first introduce a bus consisting of a poly line followed by a diff line carrying the signal gnd followed by a poly line. In the cell definition of cell zero we first define the template. In the pin section we introduce the names and the sorts of the 8 pins in, out, l.s., l.g., l.ns, r.s, r.g, r.ns of the cell, for instance, the pin in has sort metal. We then specify the distribution of the pins over the four sides of the cell and their relative position. The phrase order implicit all states that the ordering is given by the sides section, i.e. the top pin on the left side is aligned with the top pin of the right side, and so on.

An alternative to a graphical definition of cell zero would be to specify the layout within HILL. We would then replace keyword external by

```
layout
components pd: t(2,y);
begin place pd on (in right create 1, l.s.);
      route l.s.  to pd to r.s;
      route l.g   to r.g;
      route l.ns  to r. ns;
      route layer diff pd up create 1 left to in;
      route pd down to l.g.;
      route in to out;
end
```

This program assembles the layout of cell zero on a rectangular grid. This

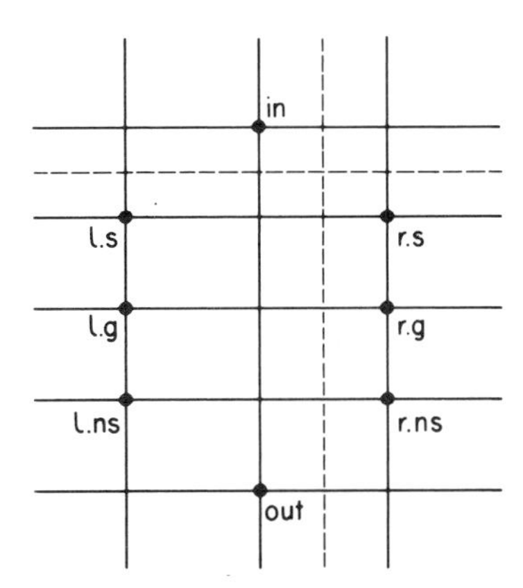

grid can grow dynamically by the use of create; it is given initially by the template (the solid grid lines in the diagram above).

In order to assemble the layout, we first declare the components needed, here a transistor pd of channel length 2 and width 4. We place pd on a new vertical grid line one to the right of in and the horizontal grid line given by l.s. Then we route a number of wires. In most of these route statements the layer of the wire can be deduced from the sort of its terminals. For example, the wire from l.s. to t must run in the poly layer, since l.s. and the gate of a transistor both exist only on the poly layer. In the next to last route statement we route a diffusion wire from the top terminal of the transistor up to a new grid line and then left to the vertical grid line determined by in. Since it hits the metal line running down from in to out there, an md contact is created automatically.

We will next describe how to build up the entire layout for the decoder from cells zero and one. Cells zero and one are to be placed into a rectangular array. For n = 3 we want to create the following pattern:

$$\begin{matrix} 0 & 1 & 0 & 1 & 0 & 1 & 0 & 1 \\ 0 & 0 & 1 & 1 & 0 & 0 & 1 & 1 \\ 0 & 0 & 0 & 0 & 1 & 1 & 1 & 1 \end{matrix}$$

In HILL there are two modes for specifying layouts. In composition mode one assembles layouts from smaller cells by abutment without explicit routing. Since HILL cells are stretchable, abutment (composition) is a method of construction that is frequently used. In layout mode one assembles the layout on a rectangular grid. Components are placed on this grid by place statements, and wires are routed between them by route statements.

In composition mode the decoder is defined by

```
cell decode (n: int) =
    temp pins in, out: array[0..2^n−1] of metal;
              ℓ,r: array[0..n−1] of bus;
         order implicit all;
         sides top in;
         bottom out;
         left ℓ;
         right r;
    pmet ;
composed
    var  i, j : int; leaf : pcell;
         begin decode := nil;
              for i from 0 to 2^n−1
              do begin leaf := nil
                   for j from 0 to n−1 do
                   begin if (i div 2^j) mod 2 = 0
                        then leaf := leaf yy zero
                        else leaf := leaf yy one
                   end;
              decode := decode xx leaf
                 end
         end
```

In this specification we build up the layout column by column. In each column we compute the column from top to bottom. After computing the type of the next cell to be abutted we add it to the partially constructed column by leaf := leaf yy zero or leaf := leaf yy one respectively.

In layout mode we assemble the layout on a square grid. The use of layout mode for the present example is less elegant than the use of composition mode; however, layout mode is a must in many cases, e.g. for leaf cells and the multiplier cell of Section 3. Even for the present example, layout mode has two advantages. First, layout mode is a "graphical" mode, and hence the system gives better error messages in layout mode; secondly, layout mode provides us with explicit names for pins of subcells which can be accessed by the simulator. For the following example we use logical to denote the common template of cells zero and one:

```
cell decode (n:int);
temp pins tin, bout : array [0..2^n−1] of poly;
          ℓb, rb : array [0..n−1] of bus;
     order implicit all;
```

```
        sides top tin;
              bottom bout;
              left ℓb;
              right rb;
pmet
layout var i, j : int;
              xstretch : xlines;
              ystretch : ylines;
        components leaf : array [0..2^n-1,0..n-1] of logical;
begin for  i := 0 to 2^n-2  do
              xstretch := tin[i] create 1;
        for  j := 0 to n-2 do
              ystretch := ℓb[i] create 1;
        xstretch := allx;
        ystretch := ally;
        for  i := 0 to 2^n-1 do
              for j := 0 to n-1 do
        begin
        place leaf [i,j] on (xstretch[2*i, 2*i+2],
                              ystretch[4*j, 4*j+4]);
        if (i div 2**j) mod 2 = 0 then
              fill leaf [i,j] with zero
        else fill leaf [i,j] with one
        end
end
```

This specification has to be interpreted as follows. We start out with a grid as given by the template (solid lines in diagram). In the components section

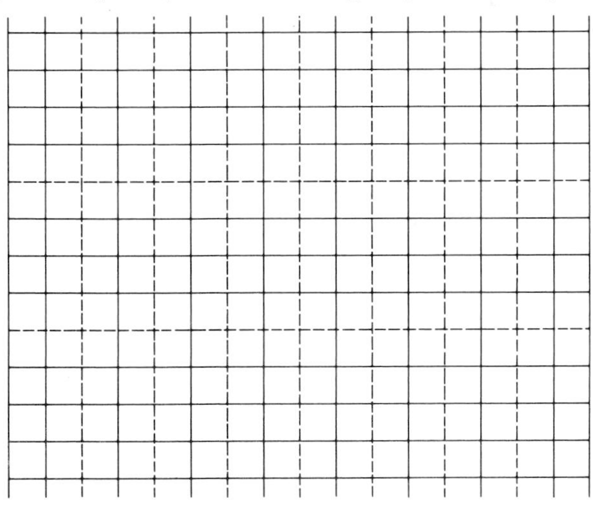

we declare an array leaf of rectangles, all of which have template logical. These rectangles are placed on the grid and then filled with cells zero or one as appropriate. In order to place the rectangles, we first create new grid lines (broken lines in the above figure), on which we can place the pins of the leaf-cell array. This is done with the create operator. Then we place the leaf cells on the grid. For each leaf cell we specify the vertical and horizontal grid lines that the template of the cell is placed upon. For this specification we use the notation *x*stretch[*i*, *j*], where *x*stretch is a list of grid lines and *x*stretch[*i*, *j*] is the sublist of *x*stretch consisting of the *i*-th to *j*-th elements. The variable all*x* and all*y* denote the list of all gridlines in the vertical and horizontal directions respectively.

In Section 5 we report about computational experience with the HILL system.

4.2 Compaction in HILL

Execution of a HILL program yields (the hierarchical representation of) a stick diagram. The compaction program takes this stick diagram and squeezes it whilst observing the design rules dictated by the fabrication process. This approach is taken by all stick-diagram-based systems [CABBAGE, FLOSS, MULGA, SLIM, STICKS, TRICKY].

Compaction is done in a number of phases $p_1, p_2, \ldots, p_k$. During the odd-numbered phases the extent of the layout in the x-direction is reduced by applying a squeeze in the horizontal direction to the layout. The y-coordinates of the layout components are not changed in odd-numbered phases. Analogously the even numbered phases squeeze the layout in the y-direction. At present, the HILL compacter treats phases independently. The reader should consult [SLIM, GW 82] for attempts to relate compaction in the x- and y-directions.

Within a phase there are two approaches to compaction. We concentrate on compaction in the x-direction in the sequel. In one approach compression ridges are run through the layout that mark areas of the layout containing excess space. This space is then removed. This process is iterated until no more compression ridges are found [MULGA, SLIM]. The second approach is graph-theoretic in nature. It is taken in HILL and also in [CABBAGE, FLOSS, STICKS, TRICKY, GW 82].

We associate a real variable with every layout component, i.e. with every wire, contact or transistor. The value of this variable represents the x-coordinate of this component. Linear inequalities (and equalities) between these variables are used to express the constraints on the x-coordinates of different layout components. There are three types of constraints.

(I) Minimum-distance constraints, $x_i - x_j \geqq a_{ij} > 0$. Constraints of this form ensure minimum separation requirements dictated by the fabrication process.

Example. Consider two parallel diffusion wires of widths w_1 and w_2. Let x_1, x_2 denote the x-coordinates of the centre lines of these wires.

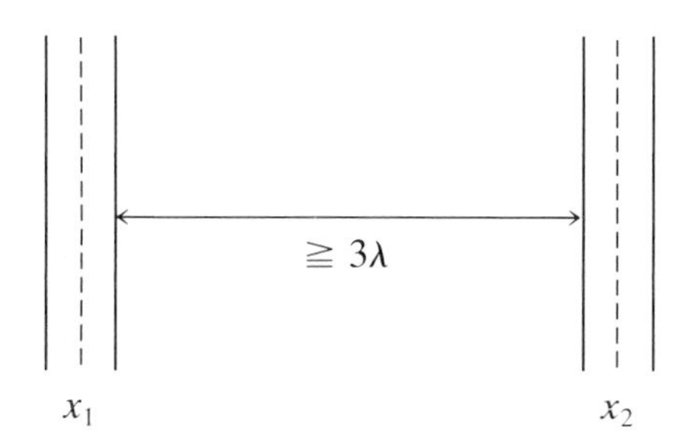

Then $x_2 \geqq x_1 + 3\lambda + (w_1 + w_2)/2$ in the process described in [MC 80].

(II) Alignment constraints, $-a_{ij} \leqq x_i - x_j \leqq a_{ij}$. Constraints of this form encode contact rules, i.e. they solder layout components together that should contact each other.

Example. Consider a square contact of side length W and a line of width w. If x_1, x_2 are the x-coordinates of the centres of the contact and the wire respectively, then

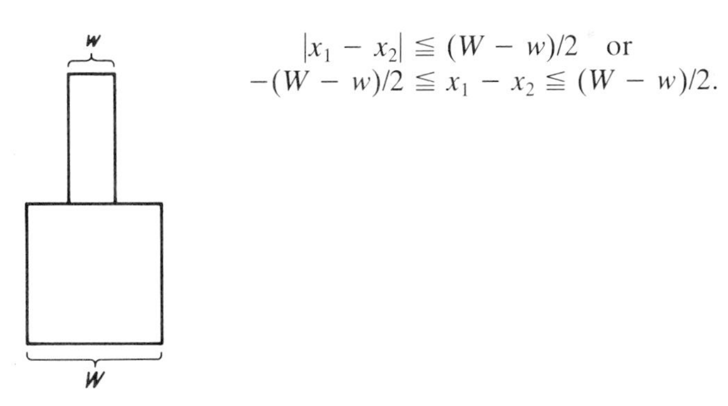

$$|x_1 - x_2| \leqq (W - w)/2 \quad \text{or}$$
$$-(W - w)/2 \leqq x_1 - x_2 \leqq (W - w)/2.$$

(III) Maximum-distance constraints, $x_i - x_j \leqq a_{ij}$. Maximum-distance constraints encode requirements given by the designer explicitly.

For example, he might wish to keep two components together because of wire delay considerations. In HILL keep-statements are used to introduce user-defined constraints.

In Subsection 4.2.1 below we discuss efficient methods for generating a sufficient set of linear inequalities. suppose now that we have a set of linear inequalities describing the layout. We then want to find values $\bar{x}_i$ for variables x_i, $i = 0, 1, \ldots$ such that $\max \bar{x}_i - \min \bar{x}_i$ is minimal (see also [Le 82a]). This problem is easily formulated as a shortest-path problem as follows. Generate a graph with nodes x_i, $i = 0, 1, \ldots$. If $x_j \leqq x_i + x_{ij}$ is a constraint (note that $x_j \geqq x_i + a_{ij}$ is equivalent to $x_i \leqq x_j - a_{ij}$, and hence all constraints are of this form) generate an edge

$$x_i \xrightarrow{a_{ij}} x_j$$

from x_i to x_j of cost a_{ij}. Also augment the graph by an additional node s and edges $s \rightarrow x_i$ from s to x_i, $i = 0, 1, \ldots$, of length 0. Then solve the single-source least-cost path problem with source s on this graph. Let $\mu(s, x)$ be the cost of a least cost path from s to x. Then $\bar{x}_i = \mu(s, x_i)$, $i = 0, 1, 2, \ldots$, is an optimal solution, i.e. a solution for which $\max \bar{x}_i - \min \bar{x}_i$ is minimal.

Example. The layout shown below might give rise to the system of inequalities $|x_1 - x_2| \leqq 1, x_3 - x_2 \geqq 2$:

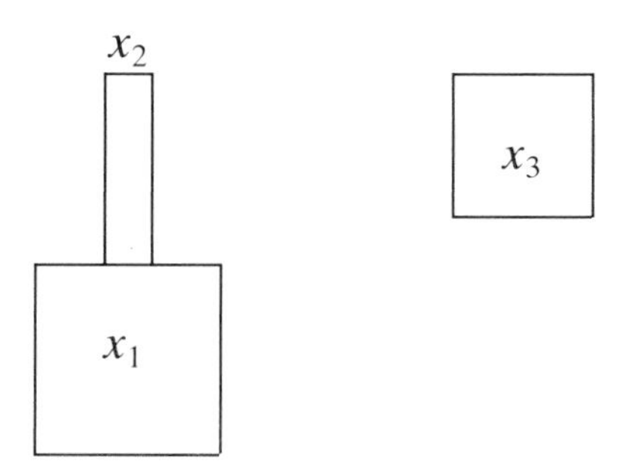

From this system we obtain the following least-cost path problem:

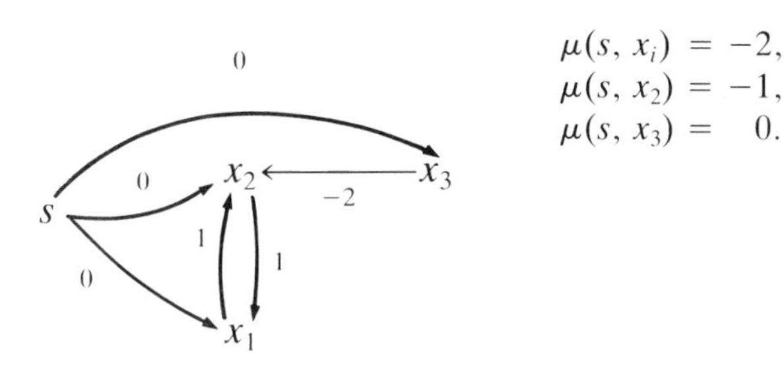

$$\begin{aligned} \mu(s, x_i) &= -2, \\ \mu(s, x_2) &= -1, \\ \mu(s, x_3) &= 0. \end{aligned}$$

Choosing $\bar{x}_1 = -2, \bar{x}_2 = -1, \bar{x}_3 = 0$ gives an optimal solution of the above system of inequalities.

In general, the correctness of this approach can be seen as follows.

(1) Note first that $\bar{x}_i = \mu(s, x_i)$ is a solution. This follows from the observation that $\mu(s, x_j) \leqq \mu(s, x_i) + a_{ij}$ whenever there is an edge of cost a_{ij} from x_i to x_j, i.e. whenever there is an inequality $x_j \leqq x_i + a_{ij}$.

(2) Suppose now that $\hat{x}_i$ is a minimal solution. We may assume w.l.o.g. that $\max \hat{x}_i = 0$. Let j be arbitrary and let p be a minimal-cost path from s to x_j, say

$$p = \xrightarrow{a_1} v_0 \xrightarrow{a_2} \dots \xrightarrow{a_k} v_k$$

with $s = v_0$, $v_k = x_j$. Then $\mu(s, x_j) = a_1 + a_2 \dots a_k$. Also $v_{i+1} \leqq v_i + a_{i+1}$ is a constraint for $0 \leqq i < h$ and hence $\hat{x}_j \leqq \hat{v}_1 + a_1 + \dots + a_k \leqq \mu(s, x_j)$. Finally, observe $\mu(s, x_i) \leqq 0$ for all i since there is an edge of cost 0 from s to x_i. Thus $\max \mu(s, x_i) \leqq 0 = \max \hat{x}_i$ and $\min \hat{x}_i \leqq \min \mu(s, x_i)$, and hence $\max \bar{x}_i - \min \bar{x}_i \leqq \max \hat{x}_i - \min \hat{x}_i$.

In Subsection 4.2.2 we discuss algorithms for solving least-cost shortest-path problems. In Subsection 4.2.3 we briefly discuss hierarchical compaction.

4.2.1 *Efficient constraint generation*

Recall that we discussed compaction in the x-direction. Efficient constraint generation is a major problem for most existing stick-diagram-based systems in a twofold sense. First, constraint generation has large running time, and secondly, it produces a large number of constraints, which in turn influences the complexity of the shortest-path algorithms used to solve the constraint system. For example, CABBAGE may produce as many as $O(n^{1.5})$ constraints, and typically produces about $O(n^{1.2})$. Of course, running time is at least that much.

Already in the HILL prototype we have overcome this deficiency and used an $O(n \log n)$ algorithm for constraint generation. The algorithm is based on a *left to right* plane sweep and used quite complicated data structures. In the spring of 1983, I. Cnop (after hearing the lectures at Louvain), R. Reischuk and Th. Lengauer independently suggested the use of a *top down* plane sweep instead. The following presentation follows [Le 83].

Note first that alignment constraints are easily generated in time $O(n)$. Also maximum-distance constraints are user-defined and therefore are of no concern here. This leaves minimum-distance constraints (Type I).

4.2.1.1 The interval graph. A straightforward way to generate the Type II constraints would be to look at each pair of layout components in turn and use the design-rule table to generate the appropriate inequality. However, this would result in $\Omega(n^2)$ inequalities—far too many to be practical.

In fact, many of these inequalities are redundant. Mostly, minimum distances are small (a few microns) such that layout components that are far apart from each other will be assured sufficient separation by the constraint that already exists in each of their neighbourhoods. Therefore only a few of the constraints are actually necessary. We will discuss how to generate such "minimal" constraint systems. To this end we formulate the following graph-theoretic problem.

Definition 1. (*a*) Let L be a set of n vertical intervals in the plane. Each interval is a triple (x, y_ℓ, y_h), where x is the x-coordinate and y_ℓ and y_h are the y-coordinates of the low and high endpoints of the interval.

(*b*) Let $(L, <)$ be the total ordering that orders L w.r.t. the x-coordinate of each interval. Ties are broken arbitrarily but fixed.

(*c*) Intervals I_1 and I_2 are said to "overlap", if

$$I_1 \,\lambda\, I_2 :\Leftrightarrow I_1 < I_2 \wedge y_{\ell,1} < y_{h,2} \wedge y_{\ell,2} < y_{h,1}.$$

(d) The following set is called the set of intervals "between" I_1 and I_2:

$$\beta(I_1, I_2) := \{I \in L;\ I_1 < I < I_2,\ I_1 \,\lambda\, I \,\lambda\, I_2\}.$$

L represents the layout geometry. Specifically, each interval represents a layout component. For a layout to be correct w.r.t. a set of design rules, we assume that overlapping intervals have to be separated by certain minimim distances in the x-direction. The exact amounts are of no concern here. Non-overlapping intervals are assumed to be sufficiently separated in the y-direction such that no constraint in the x-direction has to be generated. This can always be ensured by slightly enlarging the interval.

We shall define several ways for L to induce a so-called constraint graph $G = (L, E)$. G is a directed acyclic graph with each edge $e = (I_1, I_2)$ representing an inequality of the form $x_2 - x_1 \geqq a_{21} > 0$. Here x_1 and x_2 are the x-coordinates of I_1 and I_2 in the compacted layout. The constraint graph G is almost exactly the graph analyzed in [Le 82]. The redundancy of some constraints can now be formulated as the following axiom.

Process axiom. If $G = (L, E)$ represents an inequality system that ensures all design rules to be met—we call such a system "admissible"—then the transitive reduction ρ of G also represents such a system.

The process axiom allows us to neglect all inequalities that form "short-cuts" in the constraint graph. This is a realistic assumption because design rules are typically of a highly local nature. The process axiom is a powerful and also essentially the only existing tool for reducing the size of the set of constraints for compaction.

Clearly there are many ways of extracting a constraint graph from layout L. Here is the simplest one.

Definition 2. Let $G_0 = G_0(L) = (L, E_0)$ be defined as follows:

$$(I_1, I_2) \in E_0 \text{ iff } I_1 \lambda I_2.$$

Clearly the undirected graph underlying G_0 is an interval graph [G 80], and its interval representation is given by L, if all x-coordinates are set to zero. Thus the question of how to compute the transitive reduction π_0 of G_0 asks for algorithms to efficiently compute the transitive reduction of such interval dags. Obviously the following is true.

$$(I_1, I_2) \in \rho_0 \Leftrightarrow I_1 \lambda I_2 \wedge \beta(I_1, I_2) = \emptyset.$$

Using this characterization, ρ_0 can be computed in time $O(n \log n)$ with the following algorithm Gen$_0$. Gen$_0$ uses a top-down plane sweep. Thus the algorithm scans a horizontal sweep line across the layout L from top to bottom. During the sweep a data structure D is maintained that stores information about all intervals that currently intersect the sweep-line. D is a leaf-chained balanced tree that keeps the intervals in sorted order according to $(L, <)$. Furthermore, with each interval $I \in D$ two pointers to other intervals in L are associated. The pointer left (I) points to nil or an interval less than I. The pointer right (I) points to nil or an interval greater than I. Both pointers represent edges between I and the interval pointed to that are candidates for ρ_0 but whose membership in ρ_0 has not been decided yet. W.l.o.g. we assume that the y-coordinates for all $I \in L$ are pairwise distinct. Then during the sweep we encounter two kinds of events the algorithm has to deal with, namely the insertion of an interval into the sweep line and the deletion of an interval from the sweep line. Upon these the algorithm does the following:

```
Insert (I): Insert I into D;
            Find the left and right neighbours I_ℓ and I_r of I in d;
            left (I) := I_ℓ; right (I) := I_r;
            left (I_r) := right (I_ℓ) := I;
Delete (I): if left (I) ≠ nil and left (I) ∈ D then
               append (left (I), I) to ρ_0;
            if right (I) ≠nil and right (I) ∈ D then
               append (I, right (I)) to ρ_0;
            delete I from D;
```

Theorem 1. (*a*) Gen$_0$ computes ρ_0 in time $O(n\ log\ n)$.

(*b*) $|\rho_0| \leqq \begin{cases} 0 & \text{if } n = 1, \\ 1 & \text{if } \ = 2, \\ 2n - 4 & \text{if } n > 2. \end{cases}$

The upper bound of Theorem 1(*b*) is tight, as the following layout shows:

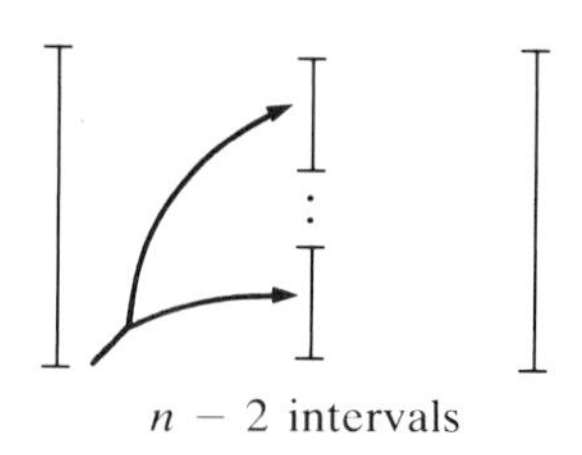

Several systems attempt to find ρ_0 [CABBAGE, SLIM]. CABBAGE may produce as many as $O(n^{1.5})$ constraints, and typically produces about $O(n^{1.2})$. SLIM comes closer, but it also produces more than the transitive closure, since a constraint is generated between each pair of intervals that are visible from each other at least in part.

4.2.1.2 The layer approach. While ρ_0 is always an admissible constraint system for compaction, it is in general not the best one. It does not allow the layout topology to be changed during compaction, because there is a constraint between an interval and all of its neighbours. [CABBAGE] states this problem without offering a solution. Within our framework we are able to generate different admissible constraint systems that entail more topological freedom. To this end we define a symmetric and reflexive binary compatibility relation $\pi \subset L \times L$. We call two intervals I_1 and I_2 compatible if $I_1 \ \pi \ I_2$. Intuitively, two intervals should be compatible if the associated components do not have to meet any minimum-distance constraint. Obviously no edge can exist in the constraint graph between any pair of compatible intervals. Therefore we make the following definition.

Definition 3. Let $G_\pi = G_\pi(L) = (L, E_\pi)$ be defined as follows:

$$(I_1, I_2) \in E_\pi :\Leftrightarrow I_1 \ \lambda \ I_2 \wedge \sim I_1 \ \pi \ I_2.$$

We make the reasonable assumption that it can be decided in time $O(1)$ if $I_1 \ \pi \ I_2$. One possibility for defining π is to realize that VLSI circuits are typically laid out on several, say l layers that are insulated from each other, except for contact holes that provide connections between the layers. Therefore we can define: $I_1 \ \pi \ I_2$ if the layout components associated with I_1 and I_2 exist on different layers that are insulated from each other. The resulting graph G_π is in general no interval dag, and its transitive reduction ρ_π may be hard to compute. But we can efficiently compute a supergraph of ρ_π with few edges.

Theorem 2. Let l be a layout such that a subset $M(I)$ of a set $\{1, \ldots, l\}$ of layers is associated with each interval $I \in L$. (We say that I exists on the layers in $M(I)$.) Assume that $|M(I) \leqq d$ for all $I \in L$. Let $I_1 \ \pi \ I_2$ iff $M(I_1) \cap M(I_2) = \emptyset$. Then we can in time $O(dn \log n)$ compute a graph R such that $\rho_\pi \subset R \subset G_\pi$ and $|R| \leqq 2dn - 4$.

[FLOSS] applies this kind of layer separation to achieve some topological freedom during compaction.

4.2.1.3 Switching the positions of components within a layer.
While ρ_π provides for more topological flexibility by handling each layer of the circuit separately, there are still desirable transformations that it does not allow. We give two examples.

Example 1. Jog-flipping of wires:

Here the intervals overlap, although slightly. Since they are on the same layer they are incompatible with respect to the above relation π and cannot exchange their positions during compaction. CABBAGE solves this problem by adjusting, specifically for such job flips, the lengths of the intervals temporarily such that they do not overlap. This solution is ad hoc, however, and it does not solve the following problem.

Example 2. Transistor flipping:

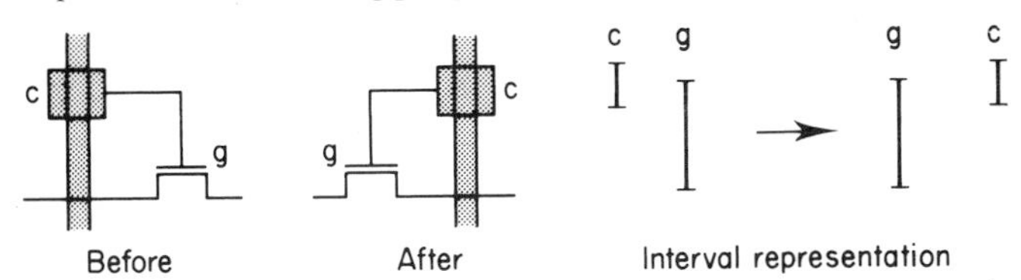

Here the vertical bold wire is on a top layer and all other structures are on the bottom layer. The contact c connects between the two layers. In this case a simple-minded adjustment of interval lengths will not do. Therefore we extend π by also allowing $I_1 \ \pi \ I_2$ if I_1 and I_2 exist on the same layer and carry the same electrical signal. Such an extension is desirable, since the above example transformations will reduce the area of many layouts significantly. But now ρ_π can become large.

Example 3.

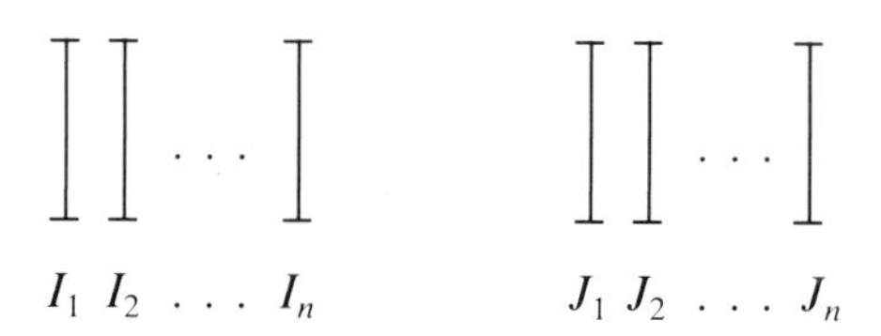

Let $I_i \; \pi \; I_j$ and $J_i \; \pi \; J_j$ for $1 \leqq i < j \leqq n$, but $\sim I_i \; \pi \; J_j$ for $1 \leqq i, j \leqq n$. Then $\rho\pi_\pi$ is the complete bipartite graph $K_{n,n}$. Indeed one can show that if one just assumes π to be reflexive and symmetric one cannot hope to find an encoding for the path information contained in G_π that has size $o(n^2)$. Thus G_π is not the appropriate constraint graph. We therefore define a new constraint graph G_1 such that $G_0 \supset G_1 \supset G_\pi$ and G_1 allows the transformations discussed above.

Definition 4. Let $G_1 = G_1(L) = (L, E_1)$ be defined as follows:

$$(I_1, I_2) \in E_1 :\Leftrightarrow I_1 \; \lambda \; I_2 \wedge (\sim I_1 \; \pi \; I_2 \vee \beta(I_1, I_2) \neq \varnothing).$$

Thus e_1 can be obtained from E_0 by deleting all edges in ρ_0 that connect compatible intervals. This allows only exchanges of the positions of neighbouring elements during compaction. However, both example transformations are included. The transitive reduction ρ_1 of G_1 can be characterized as follows.

Lemma 1.

$$\begin{aligned}(I_1, I_2) \in \rho_1 \Leftrightarrow\; & I_1 \; \lambda \; I_2 \wedge (\beta(I_1, I_2) = \varnothing \Rightarrow \sim I_1 \; \pi \; I_2) \wedge (\beta(\mathrm{I}_1, I_2) \neq \varnothing \\ & \Rightarrow \forall I \in \beta(I_1, I_2){:}\; [(I_1, I) \in \rho_0 \wedge I_1 \; \pi I] \vee \\ & \qquad [(I, I_2) \in \rho_0 \wedge I \; \pi \; I_2]).\end{aligned}$$

Lemma 1 provides the basis for an efficient algorithm for computing ρ_1. The following lemma shows that information has to be updated only locally during the algorithm.

Lemma 2. Consider an arbitrary position of a horizontal line through the layout L that does not touch the endpoints of any interval in L. Let $\rho_{1,t}$ be the subgraph that is induced from ρ_1 by all intervals intersecting the line.

(*a*) If $(I_1, I_2) \in \rho_{1,t}$ then there are at most two intervals I, I' intersecting the line such that $I_1 < I < I' < I_2$.

(*b*) The maximum in- and out-degree of any vertex in $\rho_{1,t}$ is 2.

There are examples that show that we cannot hope to find a simple one-pass algorithm for computing ρ_1 using plane sweep methods. Thus the following algorithm Gen_1 computing ρ_1 is a two-pass algorithm.

Pass 1. Run algorithm Gen_0 on L. However, output only edges $(I_1, I_2) \in \rho_0$ such that $I_1 \ \pi \ I_2$. Organize the edges in a linear list Σ of sets, each set being the collection of edges output during one delete operation.

Pass 2. Make a plane sweep bottom-up. Again maintain a balanced tree D; however, this time allow for two pointers to the left and two points to the right of each interval to store candidate edges. Furthermore, allow for one ρ_0 pointer to the left and right to store edges from Σ.

Insert(I): Insert i into D;
Fetch from the back of Σ all edges that have been output upon the deletion of i in Pass 1, and store them in the ρ_0 pointers. Maintain the set of candidate edges between the intervals at most 3 to the right or left of I in D according to Lemma 1, i.e. delete a candidate edge if the newly fetched edges from Σ show by Lemma 1 that the edge is not a candidate any more;

Delete((I): Output all candidate edges that have I as an endpoint and an interval crossing the sweep line as the other;
Delete I from D.

Theorem 3. (*a*) Gen_1 produces ρ_1 in time $o(n \log n)$.
(*b*) $|\rho_1| \leqq 4n$.

The following example shows that there are layouts such that $|\rho_1| \leqq 4n - 16$.

Example 5.

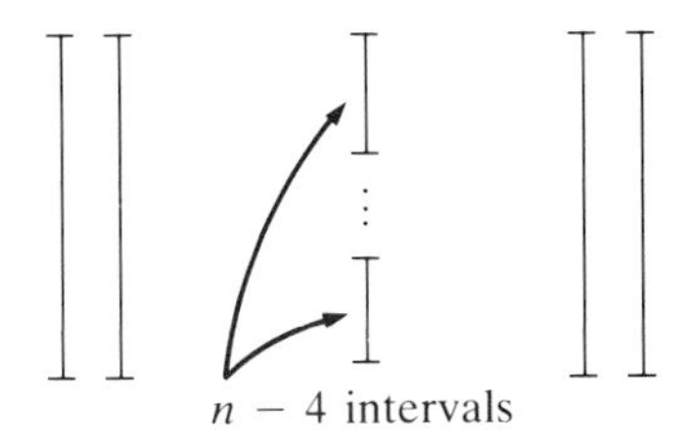

The storage requirement of both algorithms Gen_0 and Gen_1 is $O(m)$, where m is the maximum number of intervals intersecting the sweep line at any time. Since layouts can be expected to be roughly quadratic with uniform distribution of the layout components we can expect $m = O(n^{1/2})$. Here we assume that the ouptut of Pass 1 in algorithm Gen_1 is on a sequential access storage device and not in main memory. (Otherwise the storage requirement would be linear.) Both algorithms are simple enough to expect that they perform well in practice.

Additional passes can be made across the layout to generate the transitive reductions ρ_i of constraint graphs $G_i = (L, E_i)$, where

$$(I_1, I_2) \in E_i \Leftrightarrow I_1 \,\lambda\, I_2 \wedge (\sim I_1 \,\pi\, I_2 \vee \exists I \in \beta(I_1, I_2): (I_1, I), (I_1, I_2) \in E_{i-1}).$$

Then $G_0 \supset G_1 \supset G_2 \supset \ldots$. After at most n iterations the sequence stabilizes in a graph G_n with the following properties:

$$(I_1, I_2) \in E_n \Leftrightarrow I_1 \,\lambda\, I_2 \wedge (I_1, I_2) \in G_\pi^{\mathrm{T}}.$$

Here G_π^{T} is the transitive closure of G_π. Thus $\rho_n = \rho_\pi$. Unfortunately the passes to compute ρ_i become increasingly complex such that this is not a good way to compute ρ_π.

4.2.2 *Efficient constraint resolution*

Constraint resolution is tantamount to solving a single-source least-cost path problem. Least-cost path problems have received a lot of attention in the literature and are fairly well understood. A detailed discussion can be found in [M 84].

Let $G = (V, E)$ be a directed graph, $s \in V$ a special node and let $c: E \to \mathbb{R}$ be a cost function. Let $n = |V|$ and let $e = |E|$. In constraint graphs we have $e = O(n) + m$, where m is the number of user-defined constraints. Of course, m is very small in general, i.e. $m \ll n$.

Let us consider a special case first. There are no maximum distance constraints and all alignment constraints are of the form $x_i = x_j$. Then constraint graph G is acyclic, and hence the least-cost path problem can be solved in time $O(e)$ by topological sorting. This algorithm is well known and is used in [CABBAGE].

Let us now return to the general case. The best algorithm known for the general case is due to Bellman/Ford and runs in time $O(ne)$ ($=O(n^2)$ in our case). Naive use of this algorithm is out of the question because of the large size of constraint graphs. We use (or intend to use) three modifications:

(1) The constraint graph is preprocessed and its strongly connected components (SCCs) are computed. This takes time $O(e)$ by depth first search. A description of these algorithms can be found in most textbooks on algorithms. Then the least-cost path problem is solved by the Bellman/Ford algorithm on the strongly connected components and the fast algorithm for the acyclic case between components. If

SCCs are small this modification reduces running time considerably. More pecisely, the running time is $O(e + \Sigma_i n_i e_i)$, where $n_i(e_i)$ is the number of nodes (edges) in the ith SCC, $i = 0, 1, 2, \ldots$. This algorithm is implemented in the HILL prototype and works quite well.

(2) In the absence of maximum-distance constraints the SCCs are of a very special form. SCCs arise from alignment constraints between layout components that are connected vertically:

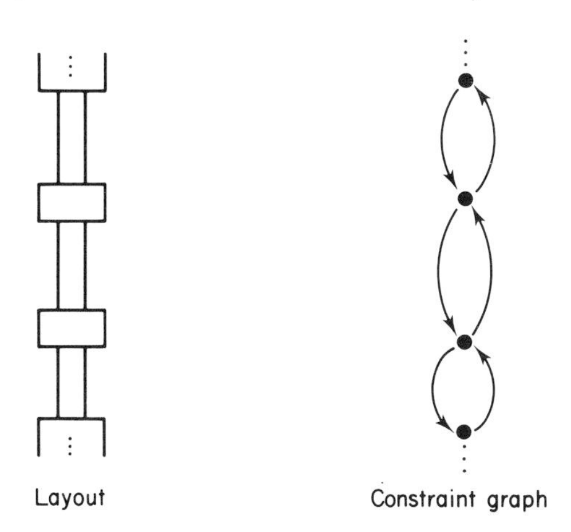

For graphs of this special form the least-cost path problem can be solved in linear time. Suppose that the nodes are numbered v_1, v_2, $v_3, \ldots, v_k$ from top to bottom and that dist $[v_i]$ is the least cost of a path from s to v_i which does not pass through any other node of the SCC shown above. Values dist $[v_i]$ are already computed when the modified algorithm reaches the SCC under consideration.

Then

> for i from 2 to k
> do dist $[v_i] \leftarrow$ min (dist $[v_i]$, dist $[v_{i-1}] + a_{i-1,i}$) od;
> for i from $k-1$ to 1
> do dist $[v_i] \leftarrow$ min (dist $[v_i]$, dist $[v_{i+1}] + a_{i+1,i}$) od

updates the dist-values correctly under consideration of the edges of the SCC. We thus have that in the absence of maximum-distance rules constraint resolution takes linear time $O(e)$.

In [Sch 83] this approach is extended to a larger class of graphs, i.e. more general (but of course not completely general) SCCs are considered and linear running time is maintained.

(3) In the presence of maximum-distance constraints the work of [MS 83] might be of some help. They show that least cost path problems on planar graphs can be solved in time $O(n^{3/2} \log n)$ instead of time $O(n^2)$ as for the Bellman/Ford algorithm. Constraint graphs as generated in Subsection 4.2 are almost planar, and hence this algorithm might be applicable.

4.2.3 Hierarchical compaction

In HILL layouts are defined hierarchically. However, no use of the hierarchy is made in the compaction process. Although we presented efficient methods for constraint generation and constraint resolution in the previous sections, the approach might run into difficulties as chips become more complex. In particular, the space requirement might become prohibitive. It is therefore worthwhile to consider hierarchical compaction. The ideas presented about hierarchical compaction are preliminary.

Suppose that we enclose every cell into a polygon with horizontal and vertical edges:

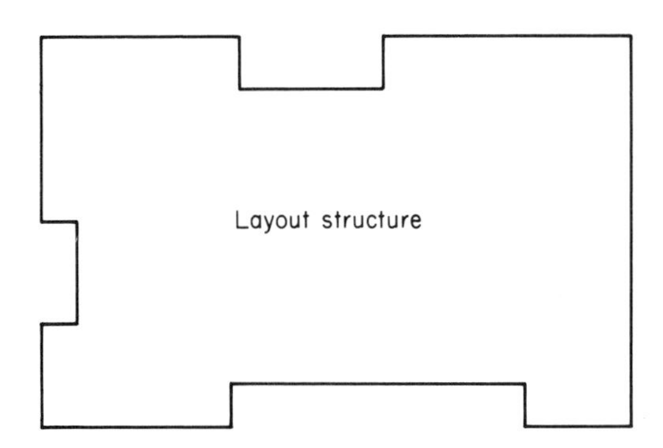

We might take as this polygon the boundary of the layout structure in the interior of the cell or a smoothing of this polygon (see below). For compaction in the x-direction we associate a variable with every vertical edge of the boundary polygon. We can now set up a set of inequalities relating the boundary and the interior of the cell.

For every occurrence of the cell as a subcell in a larger cell we conceptually replace the cell by its boundary polygon. Mathematically, we introduce a new copy of the boundary variables and use this copy of the boundary variables when we set up the inequality system for the larger cell. In general, if there are k occurrences of a cell then there are $k + 1$ occurrences of the boundary variables. One copy is used "within" the cell and one copy is used for each occurrence of the cell.

As before we can interpret the inequality system as a least cost path problem. The difference is that we deal with a hierarchically specified graph. In [Le 82b] it was shown how to solve such systems by dynamic programming. More precisely, one proceeds as follows.

(1) Phase 1 is a bottom-up phase. Starting at the leaf cells, one solves all pair shortest-path problems. When a cell is looked at, all subcells have been considered already. The subcells are replaced by complete graphs on the (copies of the) boundary variables of the subcells. The cost of an edge of the complete graph is given by the solution for the subcell.

At the end of Phase 1 an optimal solution to the constraint system has been found; more precisely, the spread (extension of the layout in the x-direction) has been found.

(2) Phase 2 is a top-down phase and actually determines the solution as follows. In phase 1 we determined the solution for the root cell. This fixes the values of the boundary variables for every occurrence of a direct subcell of the root; call such a fixing an instance of the subcell. In general, there will be more than one instance, but *less* than the number of occurrences. For every instance we add the values of the boundary variables as additional constraints to the constraint set of the cell and then solve the least-cost path problem again. This fixes the values of the boundary variables for every occurrence of a direct subcell of the root, ...

The running time of the algorithm outlined above is $\Sigma_{i=1}^{m} s_i n_i^3$, where n_i is the number of layout components in the ith cell and s_i is the number of different ways cell i is compacted. Of course, s_i might be very large.

One way of partially controlling s_i is to give the enclosing polygon a small number of edges. An extreme case is to make the enclosing polygon a rectangle. This choice guarantees a small number of boundary variables (2 + the number of pins at top and bottom), but might yield poor compaction results.

A compromise is to smooth the smallest enclosing rectangle a little but not all the way to a rectangle. For example, we might require that each edge of the polygon has length at least m units or that there are at most m edges. In the former case, [No 83] shows how to compute an optimal (= minimal-wasted-area) approximation in time $O(nm)$.

4.3 HILLSIM, a Switch-Level Simulator

Simulation at the gate level, i.e. logic simulation, has always been very popular with circuit designers. Unfortunately, it is insufficient for MOS

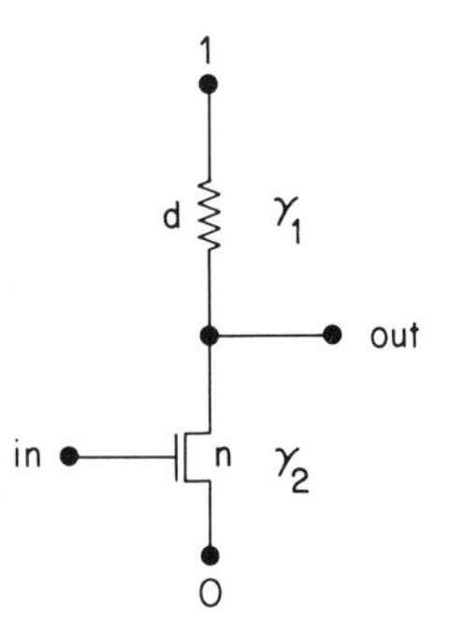

Figure 3 Inverter in NMOS.

integrated circuits because, among others, it cannot model rationed logic, dynamic memory, charge sharing and bidirectional wires. It seems that switch-level simulation plays the role of logic simulation for MOS integrated circuits. A first and very successful switch-level simulator, called MOSSIM, was developed by Bryant [Br 80]. In later papers [Br 81a, Br 81b] he gives a theoretical underpinning for the simulator, i.e. he introduces a mathematical model for the behaviour of MOS circuits and proves the correctness of the simulator with respect to the model. Unfortunately, the model is quite complicated and at some points inconsistent. Therefore, in [MNN 82] an alternative model is proposed. The model is quite different from Bryant's original model and is considerably simpler. An equivalent model has recently been proposed independently by Bryant [Br 83]. HILLSIM is correct with respect to that model and also very efficient. The simplicity of the model suggests several optimizations which could not have been obtained from Bryant's original model.

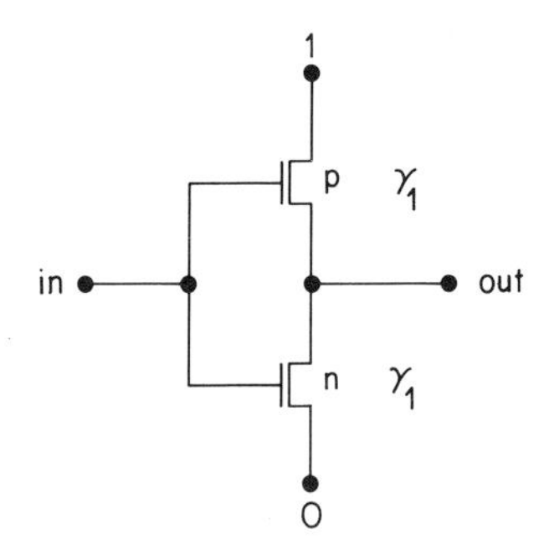

Figure 4 Inverter in CMOS.

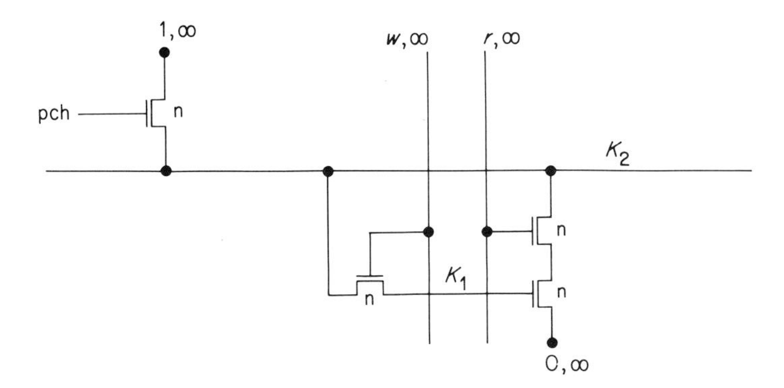

Figure 5 Memory Cell in NMOS.

We model transistors as voltage-controlled switches. An open switch has conductance 0, a closed switch has conductance γ. Here γ is an element of a finite set $\Gamma = \{\gamma_1 < \gamma_2 < \ldots < \gamma_m\}$ of possible conductances. We use three different types of transistors n, p and d. The conductances in Γ are used to model ratioed logic. We use strength (t) to denote the conductance of t in its closed state. See Figures 3 and 4.

Nodes (wires) are modelled as capacitances. Each node has a capacitance in a finite set $K = \{K_1 < K_2 < \ldots < K_q < \infty\}$ of capacitances. Input nodes have capacity ∞. The capacitances are used to model precharging and dynamic memory. We use cap(k) to denote the capacitance of node k. See Figure 5.

The state zk of a network is a function $zk: N \rightarrow \{0, 1, X\}$, where N is the set of nodes. If $zk(k) = X$ then the state of node k is either unknown (i.e. 0 or 1 but unknown) or undefined (i.e. somewhere between 0 and 1). The state of a transistor is defined by the state of the gate and the type of the transistor according to Table 2.

We can now define the basic simulation algorithm. For the sequel we use N to denote the set of nodes and T to denote the set of transistors of the network. A node state is a mapping zk: $N \rightarrow \{0, 1, X\}$ and a transistor state is a mapping zt: $T \rightarrow \{\text{open, closed}, X\}$.

Table 2.

State of gate	n	p	d
0	open	closed	closed
1	closed	open	closed
X	X	X	closed

4.3.1 *Basic simulation algorithm*

Input: a node state zk and stimuli in : $I \to \{0, 1, X\}$, where I is a subset of the set n of nodes (the input nodes);

Output: a new stable state *settle* $(zk$, in) of the network, if it exists.

Algorithm:

$$\text{Let } zk_0(k) = \begin{cases} \text{in}(k) & \text{if } k \in \text{I}, \\ zk(k) & \text{if } k \in \text{N} - \text{I}, \end{cases}$$

$i \leftarrow 0$;

repeat: let $zt: T \to \{\text{open, closed}, X\}$ be as defined by node state zk, instead of; and the table above, i.e.

$$zt(t) = \delta(\text{type } (t), zk(\text{gate } (t)),$$

where type$(t) \in \{\text{n, p, d}\}$, gate$(t) \in N$ is the gate node of transistor t and δ is given by Table 2;

$$zk_{i+1} = \text{Equ}(zk_i, zt),$$

where function Equ (Equilibrium) is defined below;

$$i \leftarrow i + 1$$

until $zk_i = zk_{i-1}$;

output zk_i.

It is important to observe at this point that the basic simulation algorithm implements a *unit-delay* assumption. Note that we first compute the transistor state as given by the node state, and then keep this state *fixed* in order to compute the equilibrium state on the nodes. Once the equilibrium is reached, we set the transistors to their new states, ... We will come back to the unit-delay assumption at the end of this section.

It remains to define the equilibrium function. We do so in a two-step process. We first define Equ(zk, zt) in the case that $zt(t) \in \{\text{open, closed}\}$ for all $t \in T$ and then extend it to arbitrary transistor states.

A transistor state zt is complete if $zt\ (t) \in \{\text{open, closed}\}$ for all transistors $t \in T$. Assume now that zt is complete. If zt is complete we define an undirected graph $G = (N, E)$ as follows. The set of vertices is the set of nodes of the network, edges correspond to closed transistors; more precisely,

$$E = \{(v, w);\ v, w \in N \text{ and there is } t \in T \text{ with } z(t) = \text{closed} \\ \text{and } \{v, w\} = \{\text{drain}(t), \text{source}(t)\}\}.$$

Let $V_1, V_2, \ldots, V_m$ be the connected components of this graph. A connected component V_i is *isolated* if $V_i \cap I = \emptyset$, i.e. if it contains no input node. For the definition of Equ we will now make a case distinction.

Isolated components. Let V_i be an isolated component. Let maxcap (V_i) = max {cap (k); $k \in V_i$} and let Equ $(zk, zt)(k) = V\{zk(v); v \in V_i$ and cap $(v) =$ maxcap $(V_i)\}$ for all $k \in V_i$, where $0 \vee 0 = 0$, $1 \vee 1 = 1$ and $0 \vee X = X \vee 0 = 0 \vee 1 = 1 \vee 1 = 1 \vee X = X$.

This definition captures the following intuition. In isolated components the nodes of maximal capacitance determine the equilibrium state. If all nodes of maximal capacitance carry the same logic value then this signal floods the entire components and X floods the component otherwise.

Example. Consider the NMOS memory cell shown above. Assume precharge = read = 0, write = 1. Then the bus of capacitance k_2 is connected with the memory cell of capacitance k_1 and these nodes form an isolated component. Hence the value on the bus is written into the cell.

Non-isolated components. Let V_i be a non-isolated component. Then every node $k \in V_i$ is connected to at least one input node by a sequence of closed transistors. The logic values carried from the input nodes along these paths will determine the equilibrium logic value of the node. We make the following definition.

Definition. A path p is a sequence $v_0 e_0 v_1 e_1 v_2 \ldots e_{k-1} v_k$ of nodes and edges such that e_i connects v_i and v_{i+1}. The strength of a path p is the minimal strength of any edge (= transistor) on the path, i.e.

$$\text{strength}(p) = \min\{\text{strength}(e);\ e \text{ is an edge of } p\}.$$

The strength of an empty path is defined as ∞. The strength of a node is the maximal strength of any path connecting it to an input node, i.e.

$$\text{strength}(v) = \max\{\text{strength}(p);\ p \text{ is a path from a node in } I \text{ to } v\}.$$

Note that this definition gives input nodes strength ∞.

Example. Consider an inverter with a pass transistor at the output (Figure 6).

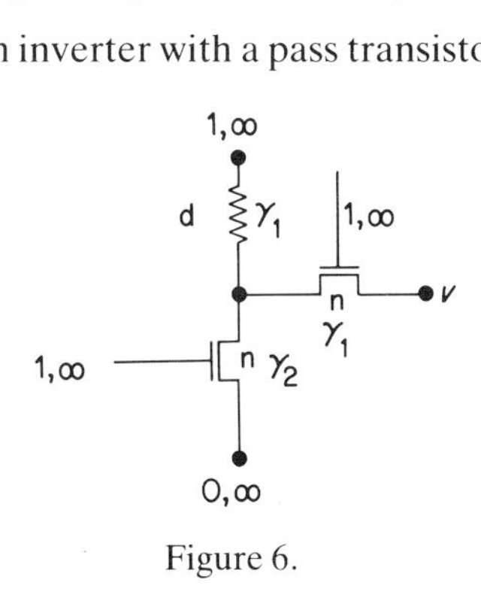

Figure 6.

There are two paths from input nodes to v, namely

both of strength γ_1. Hence v has strength γ_1.

A path $p = v_0 e_0 v_1 \ldots e_{k-1} v_k$ from an input node v_0 to a node v_k is *essential* (for v_k) iff strength $(v_i) = \min\{\text{strength } (e_0), \ldots, \text{strength } (e_{i-1})\}$ for $1 \leqq i \leqq k$, i.e. if every initial segment of p supports the strength of its end node. In our example only the second path is essential, because the output node of the inverter has strength γ_2.

The equilibrium logic value of a node is determined by the essential paths ending in that node, i.e. for all $v \in V_i$

$$\text{Equ}(zk, zt)\,(v) = V\{zk(k);\ k \in I \text{ and there is an essential path from } k \text{ to } v\}.$$

In our example above the equilibrium logic value of node v is 0. Note that it is crucial that only essential paths are considered in the definition of the equilibrium value.

The definition of equilibrium logic value given above is justified by the following.

Theorem. Let $I \subseteq N$ be a set of input nodes, let zk be a node state and let zt be a complete transistor state. Assume also that $zk(i) \in \{0, 1\}$ for $i \in I$. For $c \in \mathbb{N}$ construct the following RC-network.

Replace closed transistors of conductance γ_i by resistors of c^{-i} ohms, replace a node k of capacitance K_i by a capacitor (against ground) of c^i farads and charge it with $zk(k)\, c^i$ coulombs. Finally connect an input node i with a power supply of $zk(i)$ volts. Let voltage(c, v) be the resulting voltage at node $v \in V$. Then

$$\lim_{c \to \infty} \text{voltage}(c, v) = \begin{cases} 0 & \text{if Equ}(zk, zt)\,(v) = 0, \\ 1 & \text{if Equ}(zk, zt)\,(v) = 1, \\ x & \text{if Equ}(zk, zt)\,(v) = X, \end{cases}$$

where x is some value between 0 and 1 depending on v.

Proof. See [MMN 82].

It remains to extend the definition of Equ to incomplete transistor states. Let *zt* be a transistor state. Then *zt'* is a complete extension of *zt* if *zt'* is complete and $zt(t) \in \{\text{open, closed}\}$ implies $zt(t) = zt'(t)$, i.e. the state of undefined transistors is changed to open or closed arbitrarily. We define

$$\text{Equ}(zk, zt)\,(v) = \bigvee \{\text{Equ}(zk, zt')\,(v);\ zt' \text{ is complete extension of } zt\}.$$

This definition captures the intuition that nothing is known about the state of a transistor whose gate has logic value X. Hence all possible complete extensions have to be considered.

This completes the definition of Equ and hence finishes the description of the mathematical model of the behaviour of MOS circuits. We will now turn to the description of the simulator that *efficiently* implements this model. The main problem is to compute Equ efficiently. Again we concentrate on complete transistor states first. In this case a simple algorithm works. We explore the network, starting at the input nodes by breadth first search. All nodes encountered are entered into a set Active, from which nodes are deleted in order of decreasing strength. Hence we will first delete all nodes of strength ∞, then all nodes of strength γ_m, then all nodes of strength $\gamma_{m-1}, \ldots$ from Active. In particular, whenever we remove a node from Active we will have computed its strength. Simultaneously we propagate logic values along essential paths into the network.

In the sequel a signal s is a pair (w, st) with $w \in \{0, 1, X\}$ and $st \in \mathfrak{K} \cup \Gamma \cup \{\infty\}$. We order $\mathfrak{K} \cup \Gamma \cup \{\infty\}$ by $K_1 < \ldots < K_q < \gamma_1 < \ldots < \ldots \gamma_m < \infty$. The set $S = \{0, 1, X\} \times (\mathfrak{K} \cup \Gamma \cup \{\infty\})$ is the set of signals. Define

$\circ : \Gamma \times S \to S$ by

$$\gamma \circ (w, st) = \begin{cases} (w, st) & \text{if } st \in \mathfrak{K}, \\ (w, \min(\gamma, st)) & \text{if } st \in \Gamma \cup \{\infty\}, \end{cases}$$

and $\vee : S \times S \to S$ by

$$(w_1, st_1) \vee (w_2, st_2) = \begin{cases} (w_1, st_1) & \text{if } st_1 > st_2 \text{ or} \\ & \quad st_1 = st_2 \text{ and } w_1 = w_2, \\ (w_2, st_2) & \text{if } st_1 < st_2, \\ (X_1, st_1) & \text{if } st_1 = st_2 \text{ and } w_1 \neq w_1. \end{cases}$$

Finally, we order S by $(w_1, st_1) \leqq (w_2, st_2)$ if $st_1 < st_2$ or $(st_1 = st_2$ and $(w_1 = w_2$ or $w_2 = X))$.

The following algorithm computes Equ in the case of complete transistor states. Let *zk* be a node state, *zt* a complete transistor state and $I \subseteq N$ an input set. Then $zk' = \text{Equ}(zk, zt)$ is computed as follows:

```
(1)  for all i ∈ I do Signal [i] ← (zk (i), ∞) od;
(2)  for all k ∈ N − I do Signal [i] ← (zk (k), cap (k)) od;
(3)  Active ← K;
(4)  while Active ≠ ∅
(5)  do select k ∈ Active with maximal signal strength;
(6)     delete k from Active;
(7)     for all closed transistors t with {drain (t), source (t)}
             = {k, h} for some node h
(8)     do s ← Signal [h] ∨ strength (t) ∘ Signal [k];
(9)        if s ≠ Signal [h]
(10)       then Signal [h] ← s;
(11)            Active ← Active ∪ {h}
(12)       fi
(13)    od
(14) od
(15) for all v ∈ V do zk′(v) ← w where Signal [v] = (w, st)
                      for some st
                   od
```

We have the following.

Theorem. The algorithm above correctly computes Equ for complete transistor states. Moreover, it can be made to run in time $O(|T|)$.

Proof. For the correctness proof we refer the reader to [MMN 82]. The bound on the running time can be seen as follows. We represent the set Active by a bitvector BACTIVE $[1 \ldots n]$ with $k \in$ Active iff BACTIVE $[k]$ = true, and an array of linear lists

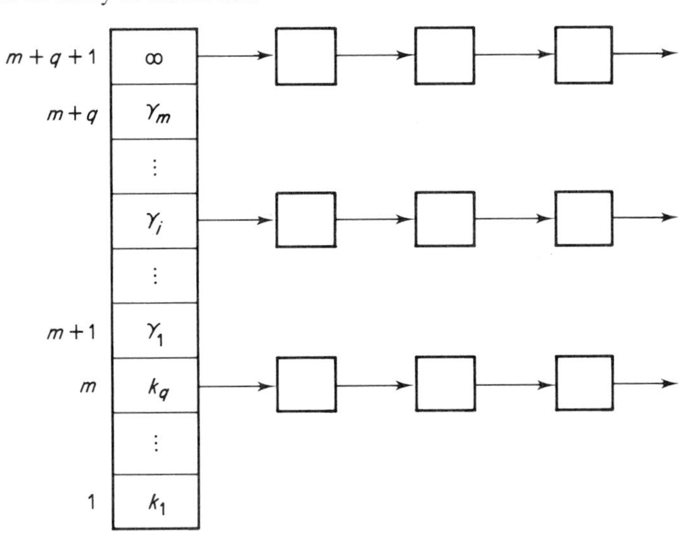

where the ith list contains all nodes in Active of strength i, a pointer max which points to the topmost non-empty list and an array of pointers of length $|K|$. If $k \in$ Active then $P[k]$ points to the location of node k in the structure of linear lists described above. With these data structures all operations on the set Active take time $O(1)$. Whenever the last element of a list is deleted we need to reset pointer max. Because max is never increased (!!) we only need to scan down the array of lists until we find the next non-empty list. Finally observe that every node is removed from Active at most twice. Thus running time is $O(|K| + |T|) = O(|T|)$ since $K \leqq 3T$. □

Extension to incomplete transistor states is quite simple. We use the algorithm above to dynamically compute two complete extensions zt^0 and zt^1 of incomplete transistor state zt. In extension zt^0 we propagate 0 and X as far as possible, and in extension zt^1 we propagate 1 and X as far as possible. We compute zt^0 and $zk^0 = \text{Equ}(zk, zt^0)$ by the algorithm described above with line (7) replaced by

(7) <u>for</u> all transistors t with $\{\text{drain}(t), \text{source}(t)\} = \{k, h\}$ for some h and either $zt(t) = \text{closed}$ or $zt(t) = X$ and $\text{Signal}[h] \in \{0, X\} \times (\Re \cup \Gamma \cup \{\infty\})$

(7a) <u>do</u> $zt^0(t) \leftarrow$ closed;

All transistors t with $zt(t) = X$ that are not explicitly closed in line (7a) are open in zt^0. Note that the algorithm above closes a transistor in the X-state only if this helps to propagate an 0 or X. A similar algorithm (replace $\{0, X\}$ by $\{1, X\}$ in line 7) is used to compute zt^1 and zk^1. Then

$$\text{Equ}(zk, zt)(v) = \begin{cases} zk^0(v) & \text{if } zk^0(v) = zk^1(v), \\ X & \text{otherwise.} \end{cases}$$

Theorem. The algorithm above correctly computes the equilibrium state. it runs in time $O(|T|)$.

Proof. See [MNN 82].

At this point we have arrived at an efficient algorithm for computing the equilibrium state. Thus one clock cycle, i.e. function settle, is simulated in time $O(h|T|)$, where h is the number of iterations required by the basic simulation algorithm. Note that each iteration takes time $O(|T|)$ by the results above. In general, h is a non-trivial number. For example, in the case of combinatorial logic h is the depth of the network.

There are several methods for improving the efficiency. We briefly describe two.

4.3.2 *Simulation in topological order*

In networks there is a natural direction in which the information flows. In

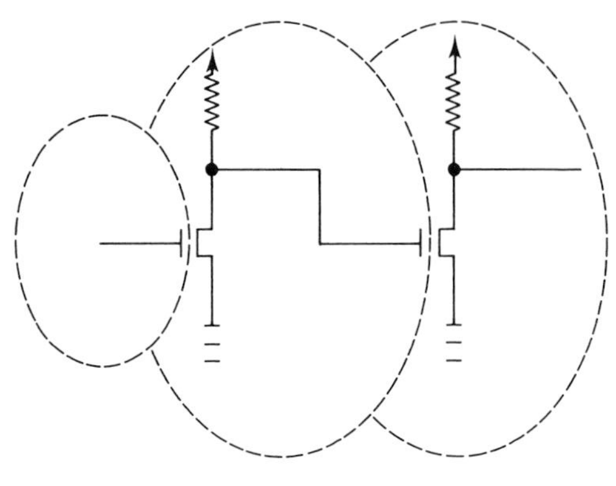

Figure 7.

particular, information always flows from the gate of a transistor to its drain and source because the state of the gate determines the state of the transistor, which in turn influences the state of its drain and source nodes. In many networks this flow of information is acyclic. We capture this idea in the following definition.

Definition. Let zk be a node state and let $I \subseteq N$. Generate a graph $g = (N, E)$ with $E = \{(v, w);\ (v, w) = (\text{drain}\ (t), \text{source}\ (t))\}$ for some transistor t and either gate $(t) \notin$ I or gate $(t) \in$ I and $\delta(\text{type}\ (t), zk\ (\text{gate}\ (t)))$ = closed, i.e. in graph G we close all transistors that are not known to be open during the entire clock cycle. Let $V_1, V_2, \ldots, V_r$ be the connected components of g. We say that V_i *influences* V_j if there is a transistor t with type $(t) \neq d$, whose gate is in V_i and whose drain and source are in V_j. Finally, we define a network to be *acyclic* (with respect to state zk and input set I) if the influence relation is acyclic.

Example. In the double inverter shown in Figure 7 we have three connected components indicated by the dashed curves. The network is acyclic.

For acyclic networks one might pursue the following strategy. Order the connected components according to the influence relation. Then simulate the first component as described above until it settles. Note that it will settle immediately, because there is no feedback within a component. Then simulate the second component, ... Simulation in topological order takes linear time for an entire clock cycle. Unfortunately, this strategy does *not* yield the same results as the simulation described above in general. The reason for this is that it uses a different timing assumption. We leave it to the

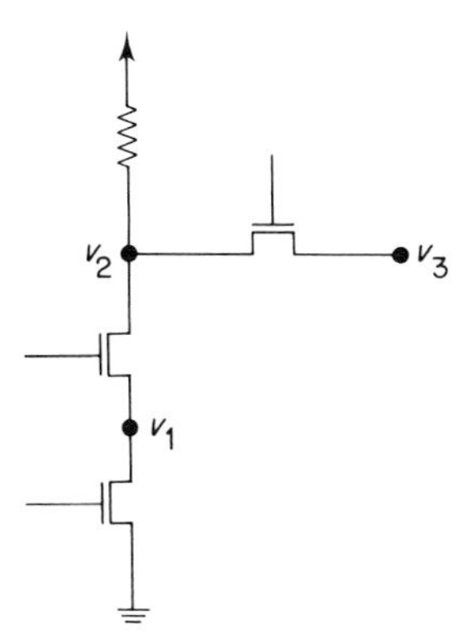

Figure 8.

reader to find a counter example. However, there is an important subclass of the acyclic networks for which the modified algorithm is equivalent to the original algorithm.

Let $0 \subseteq N$ be a set of output nodes. A node is called *strong* if it is connected to an input node by transistors of type d. Recall that type d transistors model pull-up devices. A node v is *inessential* if all paths from v to an output node or to the gate of a transistor pass through a strong node. A node v is *essential* if it is not inessential.

Example. Consider a NMOS nand-gate (Figure 8) and assume $v_3 \in 0$. Then v_2 is strong, v_3 is essential and v_1, v_2 are inessential.

Note that in a MOS-circuit every node can serve as a memory cell. We only have to isolate it from the remainder of the circuit. In particular, node v_1 in the example above can store a bit. However, the value stored in v_1 does not influence the future computation because v_1 is inessential. This observation is captured in the following Lemma.

Lemma 1. Let zk_0, zk_1 be states that agree on all input nodes and all essential nodes. Let $zk_i' = \text{settle}(zk_i, I)$, $i = 0, 1$, i.e. the network settles in state zk_i' when started in zk_i. Then $zk_0'(v) = zk_1'(v)$ for all essential nodes v.

Proof. Let zk_i^j, $j = 0, 1, 2, \ldots$, be the sequence of states computed by the basic simulation algorithm when started with zk_i, $i = 0, 1$. It is easy to show by induction on j that $zk_0^j(v) = zk_1^j(v)$ for all essential nodes v. For $j = 0$ there is nothing to show. For the induction step observe first that the components defined in the definition of equilibrium are the same because gates of transistors are controlled by essential nodes. This finishes the argument for non-isolated components. For isolated components we only have to observe that either all nodes in the component are inessential or all are essential. □

From Lemma 1 we obtain the following.

Lemma 2. Let G be an acyclic network, let zk be a node state and let $I \subseteq N$ be an input set. If in $zk' =$ settle(zk, I) there are no isolated essential nodes then settle$(zk, I)(v) =$ settle$_{\text{top}}(zk, I)(v)$ for all essential nodes, where settle$_{\text{top}}$ is computed by the modified algorithm described above (simulation in topological order).

Proof. Let $V_1, V_2, V_3, \ldots$ be the components of the control graph sorted according to relation "influences". Assume that the claim is wrong. Let i be minimal such that there is $v \in V_i$, v is essential, non-isolated and settle$(zk, I)(v) \neq$ settle$_{\text{top}}(zk, I)(v)$. Since i is minimal and all transistors between nodes in V_i are controlled by essential nodes in V_j, $j < i$, both simulations determine identical transistor states for the transistors connecting nodes in V_i. hence the same value is computed in both simulations for all non-isolated nodes in V_i, a contradiction. □

Lemma 2 tells us that simulation in topological order works correctly for a large class of networks. This class of networks includes all combinatorial networks. Sorting in topological order computes the settling state in time $O(|T|)$. This is a significant improvement over the basic algorithm.

4.3.3 Local simulation

In many cases changing the value of an input node influences only a small part of the network. This observation can be built into the simulator easily. We only have to initialize Active differently, namely to all nodes that are drain or source of a newly set transistor and all nodes reachable from these nodes by closed or undetermined transistors. For details we refer the reader to [MMN 82].

Computational experience with the simulator is quite favourable. Typically, a clock cycle takes about 0.2 ms per transistor on a Siemens 7760.

We would like to close with a short remark about the delay assumption. The basic simulator is based on the unit-delay assumption. There have been several proposals to extend switch level simulation such that propagation delays are included ([DM], [HHL]). In Lemma 2 above we went into the opposite direction. Lemma 2 above states in a certain sense that the settling state is independent of the particular propagation delays. Information of this sort could be quite important in symbolic layout systems. Note that in these systems simulation usually precedes compaction, and hence simulation has to be done without precise knowledge of delays. It would therefore be very desirable to have a simulator or a network analyser that indicates that a network is hazard-free no matter what the propagation delays are. Results in this direction are reported in [Nä 83].

5 RECENT RESULTS

Since 1983, when this series of talks was given at Louvain, several new results have been obtained. In this section we give a brief overview.

A complexity theory for VLSI: In Theorem 7 of Section 2 it was shown that the deterministic communication complexity might be almost the square of the Las Vegas complexity. Aho, Ullmann and Yannakakis [AUY] have shown that it is never more than a square. The recent paper of Vitanyi [V] discusses in great detail the question of propagation delay.

Efficient VLSI algorithms: Theorem 3 of Section 3 was improved by Mehlhorn and Preparata [MP 83]. They found a multiplier for n-bit numbers operating in time T and area $A = O((n/T)^2)$ for any T in the range $\log n \leq T \leq \sqrt{n}$. This is optimal and covers all possible values of T for AT^2-optimal multipliers.

The HILL system: The work on the system was continued. In the following we detail the present status of the system.

Layout languages: The implementation of the layout languages was completed; an up-to-date language description is available as [HILL 86]. Several small circuits (e.g., the decoder example of Section 4.1, a carry-lookahead-adder) were designed and ran successfully through the system. One large circuit (performing the scalar product of Kulisch arithmetic) is currently under development. The experience with the system shows that it is a very convenient tool at the higher level of the design. The experience also shows that the resource requirements (in particular space) are enormous for large designs. Figure 9 shows a layout produced by the system for the decoder example of Section 4.1 (including driving circuitry), $n = 4$. The current running times for this example of the various components of the HILL system are given below (VAX 780).

Compilation (preprocessor and PASCAL Compiler) 470 sec

Execution (builds up internal representation and checks its legality)	240 sec
Instantiation and Compaction	200 sec
CIF-code Generation	15 sec
Net Extraction	36 sec
Simulation (all possible input patterns)	9 sec

As the implementation of the system is optimized we expect the running times for execution, instantiation and compaction to be reduced.

Compaction: Since the paper was written more work has been done on the combinatorial aspects of constraint generation and resolution for one-dimensional layout compaction. In addition to the results of Section 3.2.1 [DL] presents a constraint generation algorithm that has a tuning parameter k. This parameter trades off area of the compacted layout with speed of the compaction. The parameter k essentially determines over how many neighbouring compatible rectangles we allow a rectangle to travel during compaction. In particular, compaction using the constraints generated with the tuneable compaction algorithm and $k = 0$ yields the same result as compaction using ρ_0. For $K = 1$ the result is similar to the result of compaction with ρ_1. For $k = n$ the compaction produces the same result as compaction using ρ_π. In general the tuneable compaction algorithm generates $2(k + 1)n - \Omega(k^2)$ constraints in time $O(n(k + \log n))$ and space $O(n)(O(\sqrt{n})$ for typical layouts). The algorithm is simple enough to be efficient in an implementation. The algorithm works only if the compatibility relation is an quivalence relation. This is not a strong restriction, however, since most often the compatibility relation is formed on the basis of signal classes.

Some progress on shortest path algorithms was also made. Mehlhorn and Schmidt [MS] define the class of BF-orderable graphs and show that the single-source shortest path problem is solvable in linear time for them. Spirakis and Tsakalidis [ST] give an algorithm for detecting negative cycles with small expected running time.

We feel that in general much progress has been made on the combinatorial aspects of layout compaction. However, the process of applying these combinatorial results, which mostly talk about rectangles on one layer with only one minimal separation distance, is still ad hoc, complicated, and error prone. It is this part of a compacter on which most time is spent. A general consistent method of making combinatorial compaction results applicable to real life layouts is still to be found.

The major weaknesses of the compacter which was used to produce Figure 9 are:

– *no jog insertion:* A very interesting proposal for jog insertion was recently made by Miller Maley [MM].

– *no wire balancing:* A heuristic for minimizing total weighted wire length is described in [SCH]. An alternative approach is to use the formulation as a linear programming problem and to customize the simplex method to it [St].

– *conservative design rules:* Our goal was to make the design of the compacter a systematic process. At times, this forces us to use more conservative design rules than the fabrication process would dictate. At present we do not see how to overcome this problem. We believe that making a systematically developed compacter into a production quality tool would be a major breakthrough.

– *interactive compaction:* Graph-based compaction allows interactive control of the compaction process. The HILL language provides a special statement, the keep-statement, for this purpose. Currently, the compacter is being made to understand keeps. One major reason for keeps is to resolve difficult interactions between the two dimensions. Recently, algorithmic approaches to two-dimensional compaction were described by Kedem and Watanabe [KW] and Schlag, Liao and Wong [SLW].

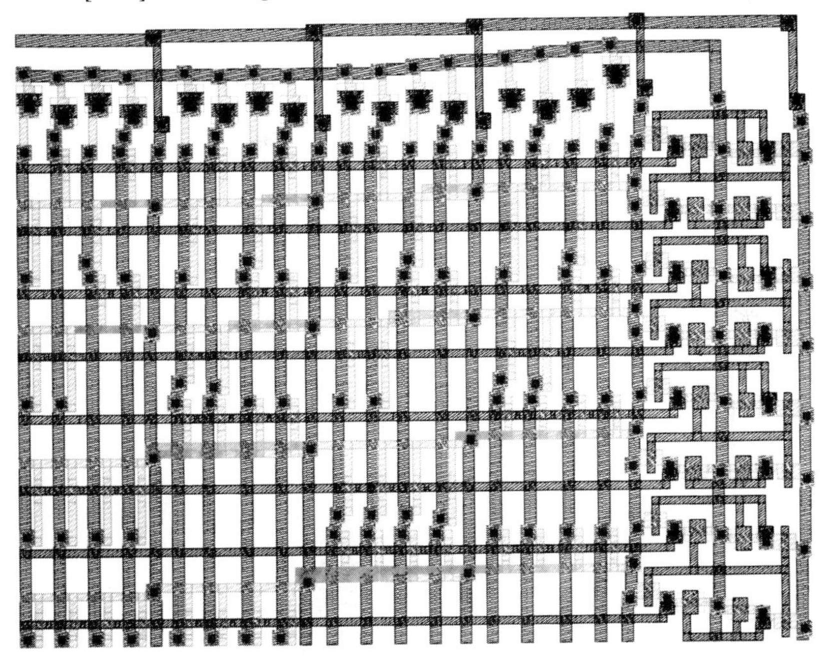

Figure 9 Layout of decoder (4).

Routing: The HILL group started work on routing algorithms in 1983. The results obtained contain an efficient switchbox router [MP 86], an efficient algorithm for local routing [KM] and a channel routing algorithm exploiting overlap [GH].

Simulation: The HILL simulator has been extended to catch races. We define a race to be a circuit node whose value depends on the internal delays inside the circuit. Being able to catch races is an important feature of a functional simulator in a symbolic design system, because at the time of simulation the exact geometries determining the internal delays in a circuit are not yet known.

The method we use to catch races is ternary simulation. Ternary simulation has been introduced by Eichelberger for Boolean gate networks as a method to catch races. It is not hard to see that ternary simulation finds every race in a Boolean gate network. The question, whether ternary simulation may also claim races at places in the circuit where in fact no races can occur, has been open for a long time. [BS] recently showed that in fact this cannot happen. Their interesting proof rests on the fact that every node in a Boolean network whose output depends on the internal circuit delays can be made to oscillate in at least one delay assignment.

Bryant [B] proposed to extend ternary simulation to MOS circuits. He supported his idea with examples. We [LN] were able to prove ternary simulation correct for MOS circuits. Our proof only uses one specific property of the HILL simulator, namely the fact that it is based on what is called a monotonic simulation model. Intuitively, a monotonic simulation model is a model in which the X value represents lack of information. Since we base our proof of ternary simulation for MOS circuits only on the monotonicity property, the result gives us considerable freedom of choice w.r.t. the simulation model. This is quite important for MOS circuits, since here technical issues such as charge sharing or voltage dividing are critical even for functional simulation. Thus different race phenomena exist, and it is very useful to the designer to have some control over which race phenomena are covered by the simulator.

Unfortunately ternary simulation does not allow for the incorporation of partial knowledge about internal delays into the simulation. Thus facts such as "node *a* always settles before node *b*" or "the delay along path *a* is always less than the delays along path *b*" cannot be taken into account. How such knowledge can be incorporated into a simulator that finds races is a major open question for Boolean gate networks as well as for MOS circuits.

ACKNOWLEDGEMENT

The authors' research was partially supported by Deutsche Forschungsgemeinschaft under Grant SFB 124, Teilprojekt B2.

REFERENCES

[AUY] A. V. Aho, I. D. Ullmann and M. Yannakakis, On notions of information transfer in VLSI circuits, *15th ACM STOC* (1983), 133–139.

[B] R. E. Bryant, Race detection in MOS circuits by ternary simulation, *VLSI* **83**, 85–98 (1983).

C[Ba 81] G.M. Baudet, On the area required by VLSI circuits. In *Proc. CMU Conf. on VLSI Systems and Computations*, pp. 100–107 (1981).

[Be 82] B. Becker, *Interner Bericht, FB 10, Univ. Saarlandes, Saarbrücken* (1982).

[Br 80] R.E. Bryant, An algorithm for MOS logic simulation. *Lamba Magazine*, pp. 46–53 (4th Quarter 1980).

[Br 81a] R.E. Bryant, A switch-level simulation model for integrated logic circuits. Ph.D. thesis, MIT; *Report* MIT/LCS/TR-259 (March 1981).

[Br 81b] R.E. Bryant, A witch-level model of MOS logic circuits. In *Proc. VLSI Conf. Edinburgh*, pp. 329–340 (1981).

[Br 83] R.E. Bryant, A switch-level model and simulator for MOS digital systems. *Caltech, CS Technical Report* 5065 (1983).

[BK 81] R.P. Brent and H.T. Kung, The area–time complexity of binary multiplication. *J. ACM* **28**, 521–534 (1981).

[BS] J. A. Brzozowski and C. J. Seger, A Characterization of ternary simulation of gate networks, *Report CS-85-37 VLSI group*, University of Waterloo, Canada (1985).

[CABBAGE] M.Y. Hsueh, Symbolic layout and compaction of integrated circuits. Ph.D. thesis, EECS Division, University of California, Berkeley (1979).

[CM 81] B. Chazelle and L. Monier, A model of computation for VLSI with related complexity results. In *Proc. 13th Annual ACM–STOC Conf.* pp. 318–325 (1981).

[DL] J. Doenhardt and T. Lengauer, Algorithmic aspects of one-dimensional layout compaction, TR 23, Fachbereich 17, Universität Paderborn (1985). (Submitted for publication.)

[DM] D. Dumlugöl and H. de Man, Logmos: A MOS transistor oriented logic simulator with assignable delays. *Technical Report, Univ. Louvain* (1983).

[E] E. B. Eichelberger, Hazard detection in combinational and sequential switching circuits, *IBM J. Res. and Dev.* 90–99 (1965).

[FLOSS] R.A. Auerbach, B.W. Lin and E.A. Elsayed, Layouts for the design of VLSI circuits. *Computer Aided Design* **13**, 271–276 (1981).

[GH] S. Gao and S. Hambrusch, Channel routing using overlap, *Algorithmica* (in press).

[GW 82] G. Kedem and H. Watanabe, Optimization techniques for IC layout and compaction. TR 117, *Dept. Comp. Sci., Univ. Rochester, N.Y.* (1982).

[G 80] M. C. Golumbic, *Algorithmic Graph Theory and Perfect Graphs*. Associated Press (1980).

[HHL] M.H. Heydeman, G.D. Hachtel and M.R. Lightner, Implementation issues of multiple delay switch level simulation. *Technical Report, Dept Electr. Engng and Comp. Sci., Univ. Colorado, Boulder.*

[HILL 84] T. Lengauer and K. Mehlhorn, The HILL System: A Design Environment for the Hierarchical Specification, Compaction, and Simulation of Integrated Circuit Layouts. In *Proc. MIT-Conference on Advanced Research in VLSI* (P. Penfield Jr., ed.) Artech House Company, 139–148 (1984).

[HILL 86] W. Rülling, Einführung in das HILL-System: Syntax, Semantik und Implementierung. *Technischer Bericht*, Universität des Saarlandes (1986).

[K 82] R. Kolla, Untere Schranken für VLSI. Diplomarbeit, FB 10, Universität des Saarlandes, Saarbrücken, West Germany (1982).

[KM] M. Kaufmann and K. Mehlhorn, Local routing of two-terminal nets is simple, *Technischer Bericht*, FB 10, Universität des Saarlandes (1985).

[KW] G. Kedem and W. Watanabe, Optimization techniques for IC layout and compaction, *Proc. 20th DA Conference* 113–120 (1984).

[Le 82a] T. Lengauer, On the solution of inequality systems relevant to IC layout. In *Proc. 8th Workshop on Graphtheoretic Methods in Computer Science* (WG 82) (Hanser, München, 1982).

[Le 82b] T. Lengauer, The complexity of compacting hierarchically specified layouts of integrated circuits. In *Proc. 23th FOCS*, pp. 358–369 (1982).

[Le 83] T. Lengauer, Efficient algorithms for the constraint generation for integrated circuit layout compaction. In *Proc. 9th Workshop on Graphtheoretic Concepts in Computer Science* (M. Nagl, ed.), pp. 139–148 (1983).

[Lu 81] W. K. Luk, A regular layout for a multiplier of $O(\log^2 N)$ time. In *Proc. CMU Conf. on VLSI Systems and Computations*, pp. 100–107 (1981).

[LM 81] T. Lengauer and K. Mehlhorn, The complexity of VLSI computations. In *Proc. CMU Conf. on VLSI Systems and Computations*, pp. 89–99 (1981).

[LN] T. Lengauer and S. Naeher, An analysis of ternary simulation as a tool for race detection in digital MOS circuits, to appear in *Integration* (Preliminary version in: VLSI: Algorithms and Architectures, Int. Workshop, Amalfi, Italy (1984)).)

[LS 81] R.J. Lipton and R. Sedgewick, Lower bounds for VLSI. In *Proc. 13th Annual ACM–STOC Conf.*, pp. 300–307 (1981).

[M 84] K. Mehlhorn, *Data Structures and Efficient Algorithms* (Springer, Berlin, 1984).

[MC 80] C. Mead and L. Conway, *Introduction to VLSI Systems* (Addison Wesley, Reading, Mass., 1980).

[MM] F. Miller Maley, Compaction with automatic log introduction, *Chapel Hill Conference on VLSI*, Computer Science Press, 261–284 (1985).

[MM 82] K. Mehlhorn and E. Meinecke-Schmidt, Las Vegas is better than determinism in VLSI and distributed computing. In *Proc. 14th Annual ACM–STOC Conf.*, pp. 330–337 (1982).

[MNN 82] K. Mehlhorn, St. Näher and M. Nowak, HILLSIM: Ein Simulator für MOS-Schaltkreise. TR A 82/08, FB 10, *Univ. Saarlandes, Saarbrücken* (1982).

[MP 83] K. Mehlhorn and F. Preparata, AT^2-optimal VLSI multipliers with minimum computation time, *Information & Control,* **58**, 137–156 (1983).

[MP 86] K. Mehlhorn and F. Preparata, Routing through a rectangle, *JACM* **33**, 60–85 (1986).

[MS] K. Mehlhorn, and B. H. Schmidt, BF-orderable graphs, *Journal of Discrete Applied Mathematics* (in press).

[MS 83] K. Mehlhorn and B. Schmidt, A single source shortest path problem for graphs with separators. In *Proc. FCT Conf. 1983*. Lecture Notes in Computer Science, Vol. **158**, pp. 302–309 (Springer, Berlin, 1983).

[MULGA] N.H.E. Weste, MULGA—An interactive symbolic layout system for the design of integrated circuits. *Bell Syst. Tech. J.* No. 60, Vol. 6 (1981), pp. 823–857.

[Nä 83] St. Näher, Diplomarbeit, Universität des Saarlandes, Saarbrücken (1983).

[No 83] M. Nowak, Diplomarbeit, Universität des Saarlandes, Saarbrücken (1983).

[PV 80] F.P. Preparata and J. Vuillemin, Area–time optimal VLSI networks for multiplying matrices. *Inf. Proc. Lett.* **11**, 77–80 (1980).

[PV 81] F.P. Preparata and J. Vuillemin, Area–time optimal VLSI networks for computing integer multiplication and discrete Fourier transform. In *Proc. 8th Int. Colloq. on Automata Theory Languages and Programming.* Lecture Notes in Computer Science, Vol. 115, pp. 29–40 (Springer, Berlin, 1981).

[SCH] W. Schiele, Improved compaction by minimized length of wires, 20th Design Automation Conference, 1983.

[Sch 83] B. Schmidt, Doktorarbeit, Universität des Saarlandes, Saarbrücken (1983).

[SLIM] A.E. Dunlop, SLIM—The translation of symbolic layouts into mask data. In *Proc. 17th Design Automation Conf.*, pp. 595–602 (IEEE, 1980).

[SLW] M. Schlag, Y. Z. Liao and C. K. Wong, An algorithm for optimal two-dimensional compaction of VLSI layouts, *Integration*, 179–209 (1983).

[St] Ch. Storb, Minimierung der gewichteten Weglänge eines Graphen, *Diplomarbeit*, FB 10, Universität des Saarlandes (1986).

[ST] P. Spirakis and A. Tsakalidis, An efficient algorithm for detecting negative cycles in a network, *JCALP* **86** (in press).

[STICKS] J.D. Williams, STICKS—A graphical compiler for high level LSI design. In *Proc. Nat. Comp. Conf.,* pp. 289–295 (1978).

[T 80] C.D. Thompson, A complexity theory for VLSI. Ph.D. thesis, Department of Computer Science, Carnegie–Mellon University (1980).

[TRICKY] A. Hanczakowski, TRICKY—Symbolic layout system for integrated circuits. *VLSI Spring Compcon* (1981).

[V] P. M. B. Vitanyi, Area penalty for sublinear signal propagation delay on a chip, *26th IEEE FOCS Symposium*, 197–207 (1985).

[V 80] J. Vuillemin, A combinatorial limit to the computing power of VLSI circuits. In *Proc. 21st Annual IEEE–FOCS Symp.*, pp. 294–300 (1980).

[V 83] J. Vuillemin, A very fast multiplication algorithm for VLSI implementation. *INRIA Report* 183 (1983).

[Y 79] A.C. Yao, Some complexity questions related to distributive computing. In *Proc. 11th Annual ACM–STOC Conf.*, pp. 209–213 (1979).

[Y 81] A.C. Yao, The entropic limitations on VLSI computations. In *Proc. 13th Annual ACM–STOC Conf.*, pp. 308–311 (1981).

3. Statistical properties and layout strategies for NMOS and CMOS layout

F. ANCEAU

The design of the layout of complex ICs at the level of quality which is necessary for mass production needs to follow some topological rules about the organization of their floor plan. The shape of the blocks must be adjusted to their environment by using their deformability function. A general deformability function which may be parameterized to each kind of functional block will be presented. The wiring strategies of the blocks come from their organization by the crossing of two straight data and control systems. The allocation of these streams to conductive layers depends strongly on the features of the technologies used. The use of NMOS and CMOS simple or double poly/metal layers will be considered. The organization of the layout by trips of overlapped logic cells and connections will be presented as a good engineering solution. These results allow the prediction of the organization of the floor plan of future VLSI from their functional specifications.

1 INTRODUCTION

The optimization of the layout of VLSI (see Figure 1) designed for mass production follows some simple rules.

(i) The shape of blocks that constitute a VLSI must be adjusted to their destination place. The modification of the shape of blocks is controlled by their deformability function.

(ii) Block interconnections must be optimized by:

 (*a*) designing blocks which may be directly "plugged" together (abutment);

 (*b*) passing-through a block by a connection instead of turning around it.—the availability of a block to be crossed by a connection without any dimensional change is given by its "transparency" functions per direction and per conductive layer:

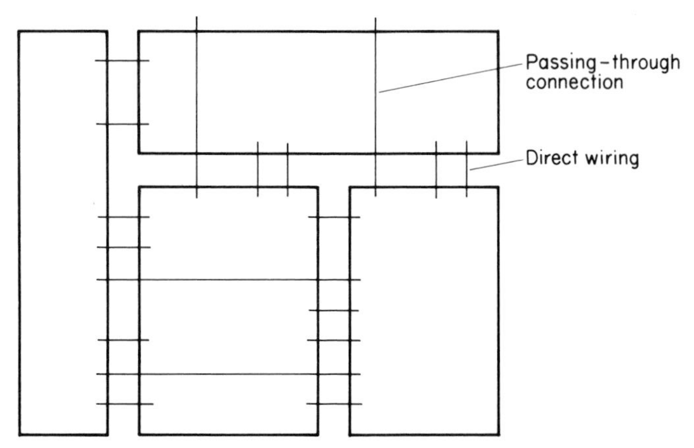

Figure 1 Typical Organization of Blocks and Connections in a VLSI Circuit.

(*c*) using the space under the connections to put the transistors—experimental and theoretical results [1] show that in VLSI circuits the area occupied by the connections is greater than, or equal to, the area occupied by the functional cells.

Obviously, following these rules leads to the drawing of a larger layout for the cells than their minimum possible area (by approximately a factor of 1.5). But the total area for the IC is reduced from 2 (1 for cells and 1 for routing) to 1.5 (larger cells under wires), which provides roughly an area reduction factor of 0.25.

This layout strategy means that it is better to consider global optimization than local (cells) optimization. This is strongly linked to the structured approach for VLSI design by looking for the topological relationship between blocks at every level of the top-down approach.

The drawbacks of this layout strategy are as follows:

(i) A very low rate for the reusing of the layout of previously designed cells. The adaptation of the layout of a block to its environment makes it specific to a given application. The use of program to generate layout of blocks overcomes this drawback because these routines may be very general and reusable.

(ii) The speed of a block may be decreased by the use of a non-optimal internal organization which is used to give it a requested shape. This is a trade-off situation where the designer uses his Art to balance a circuit between speed and silicon area. From a practical point of view, all the blocks of a VLSI do not need to be optimized in the same way (speed or silicon area).

Applying these rules implies the use of a floor plan to manage and control the design of a whole VLSI circuit. The floor plan is a data structure which allows the solving of topological and connection problems before the layout of the VLSI circuit will be completed. Drawing a floor plan requires prediction of the area and the shape of the main blocks and their connections with a reasonable accuracy [2].

2 CELL DEFORMABILITY

The prediction of the area, shape and connections of a cell need the selection of an internal organization which satisfies these requirements. Experiments show that the general deformability function of the layout of a block is given by

$$y = \frac{K}{x - \text{delt}} - \text{delt},$$

where "delt" is the minimum size of a block and K is a coefficient linked to its minimum area. The value of delt is linked to the minimum size of elementary devices (transistors, contacts) like a "granulometry" of the layout, and is also linked to the maximum levels of wiring in each direction. The internal organizaton and the area of a block vary when its shape varies (Figure 2).

3 WIRING STRATEGY

The wiring strategy of a VLSI is the technique for using as efficiently as possible the conductive layers that technology provides. These conductive layers may be considered in decreasing order of their availability to transmit electrical signals (metal, silicide, polysilicon, diffusion). The main problem to be solved by the wiring strategy is the use of these layers to make the connections according to their length.

Unfortunately the practical number of conductive layers a technology provides that may be used to realize the connections is small, because the first polysilicon layer and the diffusion layer are already used to make transistors.

The layout of the main blocks of a VLSI is designed either by using well-known techniques elaborated for some specific blocks (ROM, RAM, PLA) or by using a general strategy (random logic), which consists in organizing the block as the crossing area of two data and control streams

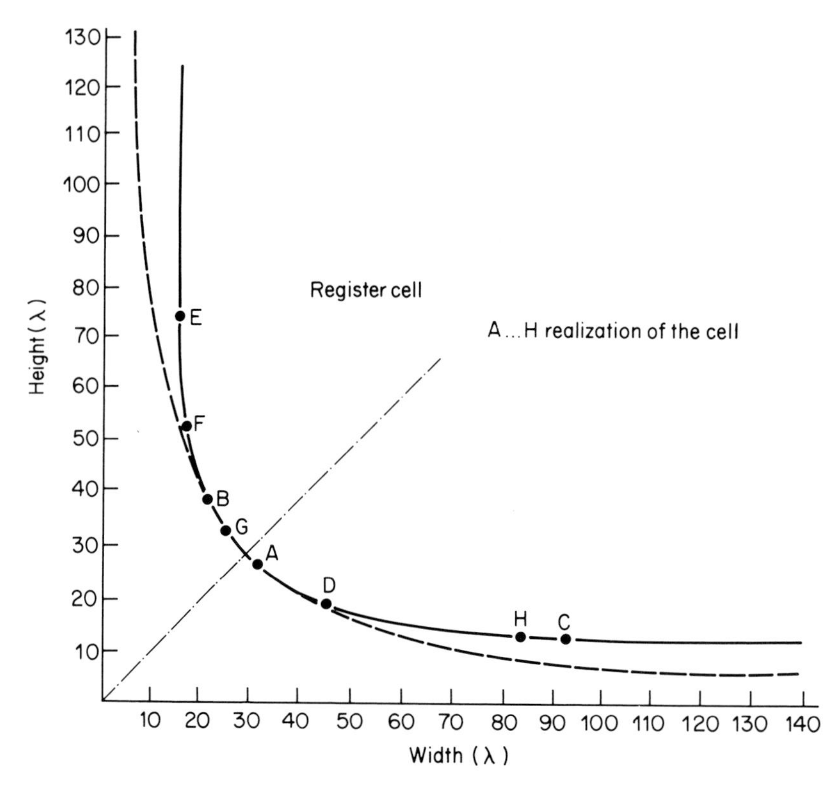

Figure 2 Deformability Curve of a Layout Block.

(Figure 3). These streams are implemented by using the two best conductive layers that technology provides.

This strategy leads to the organization of the block as an array of cells located at the crossing of each data wire and each control wire. This approach eliminates internal routing and reduces design cost. This way of designing a block from the position of its connections at its boundaries is very suitable for providing a cell layout that may be directly connected to the cell environment without the need for any routing area.

The transparency of the cells to these conductive layers determines their wiring organization (Figure 4). A high transparency (e.g. for the metal layer) allows the implementation of the corresponding stream as a set of straight wires running above the cells. A low transparency (e.g. for the polysilicon layer used both for making transistors and wiring) requests the implementation of the corresponding stream into routing channels between cells.

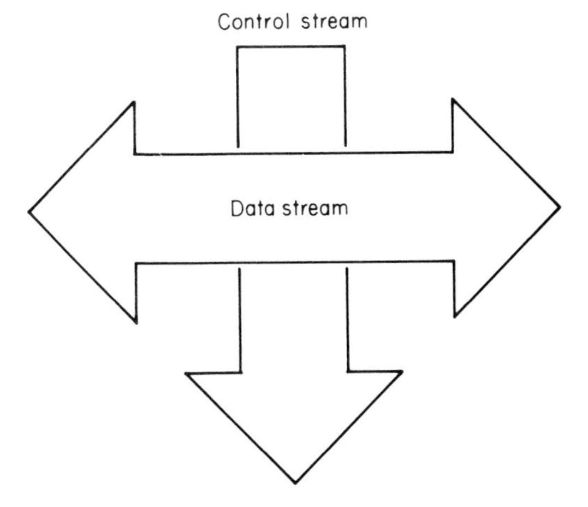

Figure 3 Data and Control Streams.

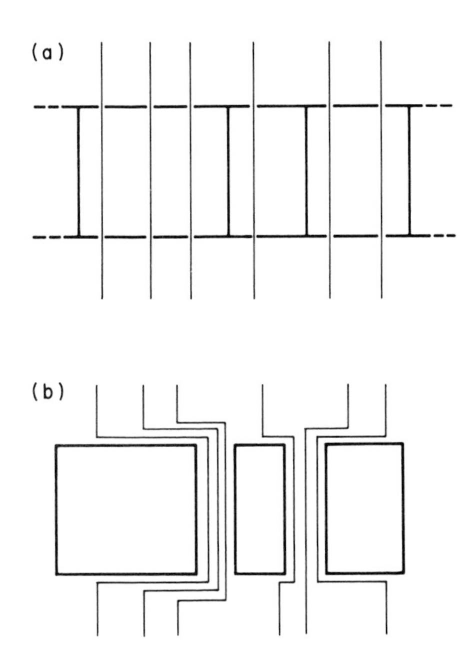

Figure 4 Wiring Strategy Depending Upon Cell Transparency: (a) High Transparency; (b) Low Transparency.

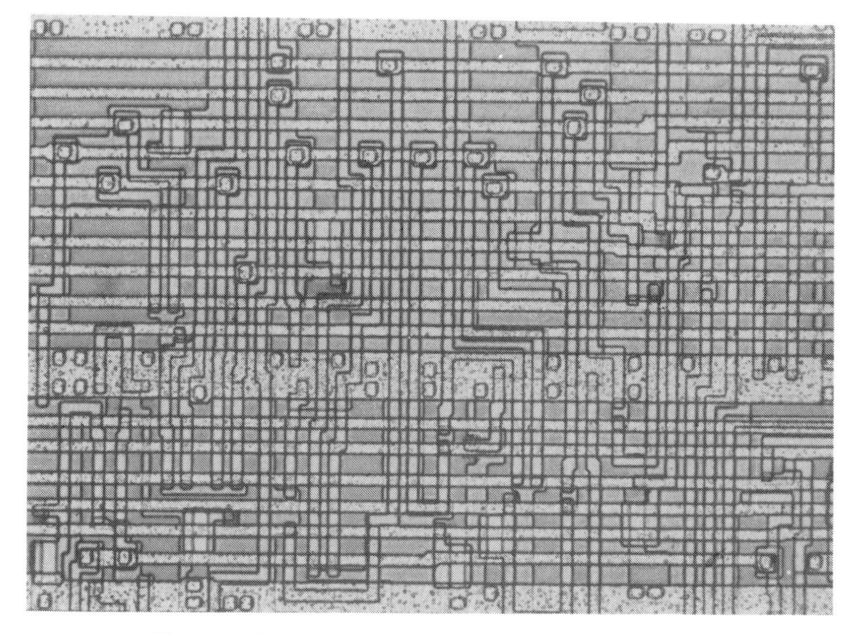

Figure 5 Strip Layout (From MC 6800 LSI Processor).

4 NMOS/HMOS TECHNOLOGIES

The dissemblance of cell transparency between the metal and polysilicon layers of the NMOS/HMOS technologies leads us to organize the layout for random blocks as strips of cells located between the teeth of the metallic power supply combs (Figure 5). The straight metallic wires, which implement a stream, run parallel with the power supply lines, above the cells, which are transparent for this layer. The polysilicon wires that implement the other stream are located into routing channels between cells, which are not transparent for this layer.

The resolution of the topological problems in cell layout (e.g. feedback connections) and some technological features (e.g. butting contacts) introduces perturbations in the organization of the upper conductive layers which should be fully allocated to the realization of a stream. These perturbations are small for NMOS/HMOS technologies, but higher for CMOS technologies.

4.1 NMOS Double-Poly and Double-Metal Technologies

These technologies provide an extra conductive layer, which is used to implement the second stream without being perturbed by making gates. The

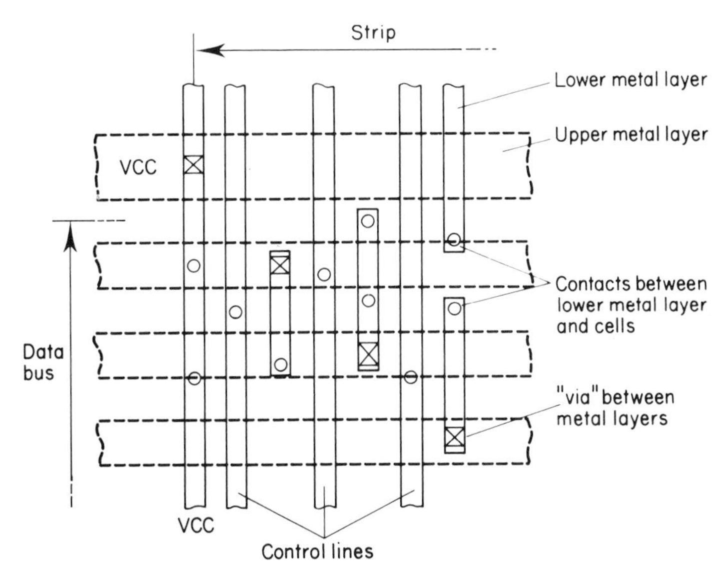

Figure 6 Double-Metal NMOS/HMOS Layout (HP 9000).

efficiency of this extra layer depends strongly on its connection capabilities. In every case the connections between the upper conductive layer and gates introduce perturbations in the intermediate conductive layer (holes or relays).

Double-poly technologies

The added polysilicon layer may be only connected to a metal layer, which is alone allowed to be connected to every layer. The connections between the second polysilicon layer and the cells must use metal bridges (butting contacts). These bridges introduce perturbations into the metal layer. In order to reduce this effect, the stream that is less connected to cells is implemented using the second polysilicon layer.

Double-metal technologies (*Figure 6*)

The added metal layer is put above the normal metal layer. This new layer may only be connected to the normal metal layer by using a new kind of contact called "via". The two streams are implemented by using the two metal layers. Because the deepest one may be connected to every layer, it implements the stream that is mostly connected to the cell elements. The upper metal layer implements the stream that is less connected to cell elements (and main power distribution). This stream is often that which has the maximum span. The connections between the upper metal layer and cells introduces perturbations in the lower metal layer as relay strips. The layout is more efficient than those using double polysilicon. It is organized as a strip in the direction of the deepest metal layer as in the case of single-poly and single-metal technologies. The strips defined by the upper metal layer have few relationships with cell layout.

5 TRANSPARENCY FUNCTION

The transparency T of a set of cells in a strip, for the metal wires, in NMOS technology (single-metal and single-poly) is given by a decreasing exponential function which indicates how many wires may manage to pass through this set of cells (Figure 7):

$$T = 0.7\,\frac{Y_m}{p_m}\,e^{-X_p/20p_p},$$

where X_p and Y_m indicate the block sizes in the poly and metal directions, and p_p and p_m are the polysilicon and metal pitches (including alternate contacts). Statistical study of the layout of commercial microprocessors [2] has shown that the mean number of logical metal wires in a strip is between 6 and 8. This value is used to estimate the strip pitch during the floor-plan prediction step.

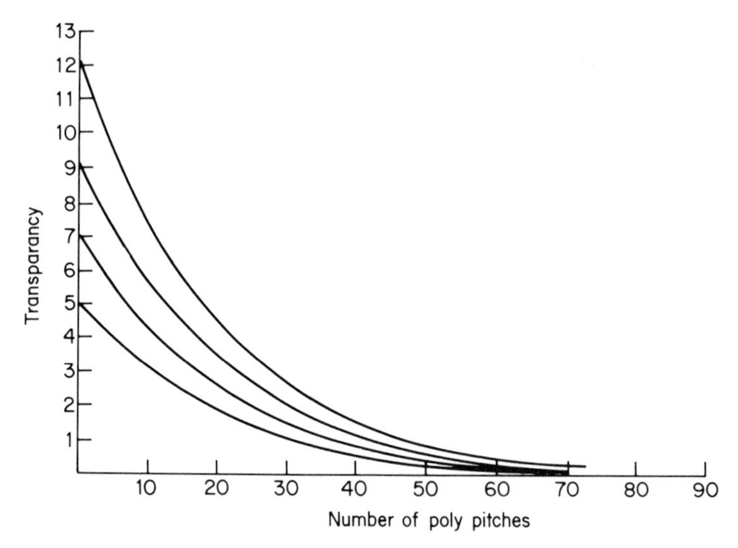

Figure 7 Transparency Curves of a Strip of Cells.

6 LAYOUT DENSITY FOR TRANSISTORS

The same statistical study has shown that the product of polysilicon and metal pitches (measured with alternate contacts) is a good unit to measure the layout density for transistors for several NMOS/HMOS technologies.

The mean layout area per transistor is close to $4p_p \cdot p_m$ for hand-optimized cell layout and $13p_p \cdot p_m$ for a large layout area. Large and regular structures that obey the optimization rules (e.g. a data path) have a mean layout area of $6p_p \cdot p_m$ per transistor.

7 CMOS TECHNOLOGIES

The use of CMOS technology introduces new constraints as follows.

(i) Static circuitry requires that the gate of each n-transistor must be connected to a gate of a p-transistor. This feature increases the topological problems to layout gates.

(ii) The polysilicon layer (n-doped) cannot be directly connected to p^+ diffusion without making a parasitic diode.

7.1 CMOS Single-Poly and Single-Metal Technologies

Using these technologies brings us to the point of organizing each CMOS logical gate as two facing diffusion strips (n^+ and p^+) crossed by common polysilicon strips making the gates of the p- and n-strip transistors (Figure 8).

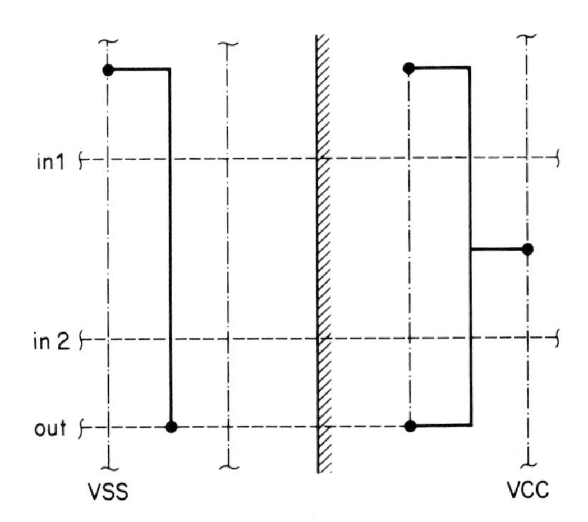

Figure 8 CMOS Single-Poly Single-Metal Basic NAND Gate.

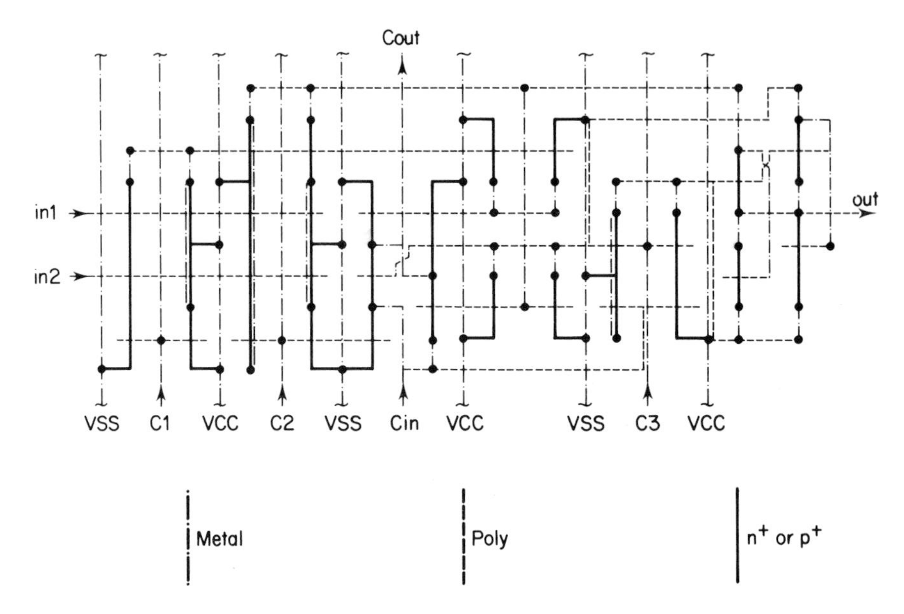

Figure 9 CMOS Single-Poly Single-Metal Layout.

Every static gate (except the inverter) contains transistors in parallel, which are obtained by making strap(s) on one of the diffusion strips. The alignment of each type of diffusion strip requires that the teeth of the power supply combs be parallel to these diffusion strips to feed each type of transistor. Wells are also organized as large strips in the same direction.

The layout is also organized in the form of strips. Metal lines run parallel to diffusion strips between the teeth of metal power supply combs and are perpendicular to poly lines. The straps used to put transistors in parallel are also made of metal because they run in the same direction. The n- and p-transistor networks of each gate are connected by a poly strip, which may be connected to the n^+ diffusion area by a buried contact (if allowed) and to the p^+ diffusion area by a butting contact. The use of metal strips to close gates and put transistors in parallel introduces a metal *L* shape which eliminates gate transparency in both directions. The use of a butting contact to connect poly trap to p^+ diffusion suggests the use of NAND gates for logic because their p-transistors are put in parallel by a metal wire directly connected to this butting contact using the same layout track. Electrical considerations about mobility also suggest the using of NAND gates.

A first approximation indicates that the silicon area occupied by a CMOS single-poly single-metal layout (Figure 9) is about 1.5 larger than a functionally equivalent NMOS/HMOS single-poly single-metal layout.

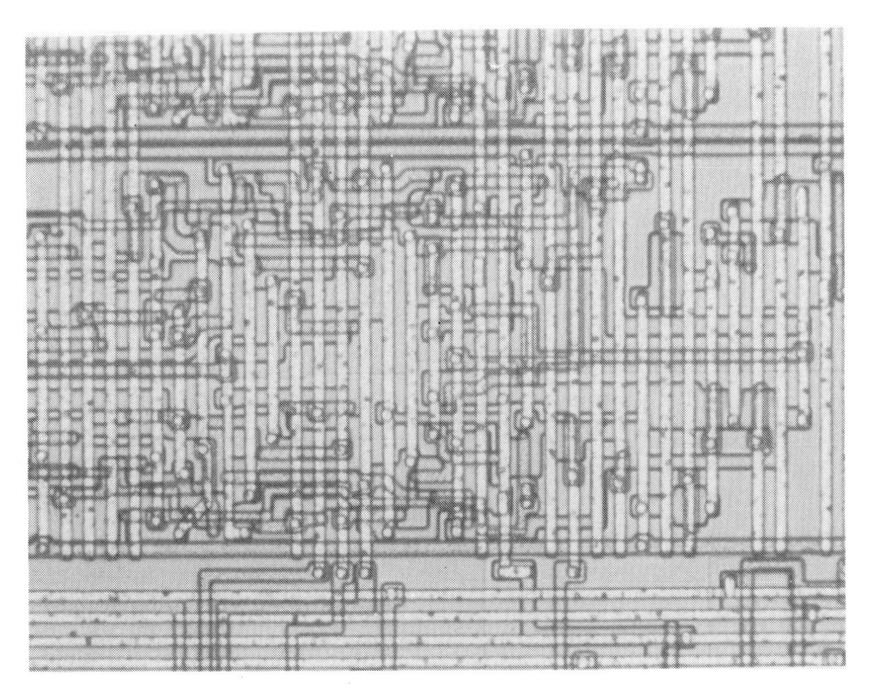

Figure 10 CMOS Double-Poly Layout (From NSC 800 LSI Processor).

7.2 CMOS Double-Poly and Double-Metal Technologies

The use of a second polysilicon layer allows the use of the first polysilicon layer only to make transistor gates (which may be independent). This eliminates the topological problems of CMOS single poly, and allows the use of diffusion straps to put transistors in parallel. The two pure conductive layers are used to implement the data and control streams. The output strap of a gate connecting n^+ and p^+ diffusion area is made with the metal layer, which may be directly connected to these booth diffusions. Then this wiring layer becomes parallel to the gates of the transistors and is used to connect n- and p-transistor networks. The other wiring layer and local power supply must be implemented by the second polysilicon layer. The low power consumption of static gates allows them to be fed locally by poly wires. The main distribution of power supply is made with the upper metal layer. (See Figure 10.) In the case of the use of two metal layers, it seems better to organize the layout as in the case of single-poly and single-metal technology, where the intermediate metal layer plays the role of the single-metal layer. As in NMOS/HMOS technologies, straps in this intermediate layer must be added as relays to connect cells to the upper metal layer.

A first approximation indicates that the silicon area occupied by a CMOS double-metal design is about the same as a functionally equivalent NMOS/HMOS single-poly single-metal design.

8 A TOOL FOR FLOOR PLANNING

The design of the floor plan of a VLSI must be made with an accuracy level depending upon the advance of the design process for each block. These levels are as follows.

(i) A first level of estimation of the size and the shape of the main blocks is useful during the preliminary stage of floor planning in order to create the main topological structure of the chip. This first estimation level is based on the estimation of the block complexity. The shape of the block may be adjusted to its destination place by using the standard deformability function presented in Section 2. At this level the connections are given as a set of equipotential names connected to the ports of the blocks.

(ii) A more accurate second level of estimation is based on the choice of internal organizations for the blocks. Each block organization gives a possible location of its ports. This second level is used to improve the floor-plan accuracy and extend the topological structure inside the blocks. At this level a prerouting of the connections consists in the allocation of the equipotential to the routing channels, pass-through wires, etc. This information allows estimation of the size of the routing channels and the change in the size of the blocks due to the connections passing through them by using the transparency function presented in Section 5.

(iii) The third level consists of the real layout of the blocks and connections.

These three levels are used to progressively create and extend the chip structure. The tool we are developing is called FLOPE–TESS [3]. It consists of three main parts:

(i) a data structure representing
  (*a*) the topological relationship between the blocks,
  (*b*) the hierarchical structure of the imbrication of the blocks,
  (*c*) the functional structure of the connections;

(ii) a graphical editor used to dialogue with the designer for
  (*a*) moving, turning, mirroring the blocks,

(*b*) getting the topological constraints the designer applies to the blocks,
(*c*) getting the information about the pre-routing of the connections;

(iii) a set of estimation routines consisting of
(*a*) the standard routines for size and deformability prediction,
(*b*) the specific evaluation routines which decide the suitable internal organization of each type of block.

9 CONCLUSION

The layout strategy that is presented here is an extension of the engineering approach used to layout commercial LSI. Its application gives a significant silicon-area reduction and a decreasing of layout design cost by decreasing the amount of connection to route the layout blocks, and by an increasing of the layout regularity. The knowledge of statistical results allows the prediction of the size and shape of the main blocks of a VLSI circuit in order to build progressively its floor plan.

REFERENCES

[1] W.R. Heller, W.F. Mikhail and W.E. Donath, Prediction of wiring space requirement for VLSI. *J. Design Automation and Fault-Tolerant Computing*, **X**, 117–144 (1978).
[2] F. Anceau and R. Reis, Complex integrated circuit design strategy. *IEEE J. Solid-State Circuits*, **17**, 458–464 (1982).
[3] R. Reis, TESS evaluateur topologique predictif pour la génération automatique des plans de masse de circuits VLSI. Thèse de docteur ingénieur (1983).
[4] F. Anceau, Layout strategies for NMOS–CMOS VLSI. *ICCD Port Chester* (October 1983).

4. Silicon compilation for microprocessor-like VLSI

F. ANCEAU

This paper presents a silicon compiler for VLSI circuits organized as special-purpose microprocessors. An internal organization of an IC which looks like that of a microprocessor is a good engineering solution for implementing a large set of custom-oriented VLSIs. The design method that is implemented by the compiler is based on the use of predefined architectural templets which presolve the layout problems. The efficiency of this design method has been proved by its application in the design of several industrial ICs.

1 INTRODUCTION

A great deal of knowledge has been accumulated about the design of microprocessors. This knowledge has reached the level where it allows efficient automatic design of this kind of IC. A large set of custom-oriented digital ICs may be organized as special-purpose microprocessors and then automatically designed. The ICs of this set are defined by their expected behaviour described by an algorithm written in a Register Transfer Language.

Some research works suggest that von Neumann microprocessor structure should be the most efficient way of executing an algorithm using the available technology. The von Neumann processor structures could be the next set of components that will be used to build very large integrated circuits (several millions of transistors). The future computer organizations (data flow, oriented to logic programming, SIMD, ...) will be constituted with subunits which will often be organized as von Neumann processors.

These considerations show the interest in designing, rapidly and efficiently, von Neumann processor structures.

2 GENERAL CONSIDERATIONS

The design methodology we want automatized is based on the use of predefined architectural templets which record good solutions. These templets presolve the problems encountered during the design of microprocessors. These good solutions represent, in fact, good trade-off between contradictory design parameters.

These templets come from the engineering world, where a lot of engineers try to obtain the most efficient solutions taking account of the available technology. Formalization and extension is necessary in order to transform these good engineering solutions into usable templets.

The design of a microprocessor-like VLSI may be divided into two main steps, which are

(i) data-path design step;
(ii) control-section design step.

The initial specification of the IC is given by an algorithm that describes what the IC must exactly perform step by step. The actions (e.g. transferts between registers) invoked by this algorithm are realized by the data path. The control structure of the algorithm is performed by the control section. The first design operation (after the algorithm has been debugged) is to split this algorithm into

(i) the set of action instructions it invokes;
(ii) its control structure.

This splitting operation may be performed by a syntactic analyser of this description language, where routines are embedded to route each recognized input string to one of these classes. Another program is also used to keep only one occurrence of each action instruction and remove conditions selecting transferts or embedded into expressions.

3 DATA-PATH DESIGN

3.1 Data-Path Templets

The best templet that may be used to design data paths comes from the MC 68000 microprocessor. This templet provides models for

(i) functional organization of the data path;
(ii) electrical behaviour of the buses of the data path;
(iii) topological organization of the layout of the data path;
(iv) testability of the data path.

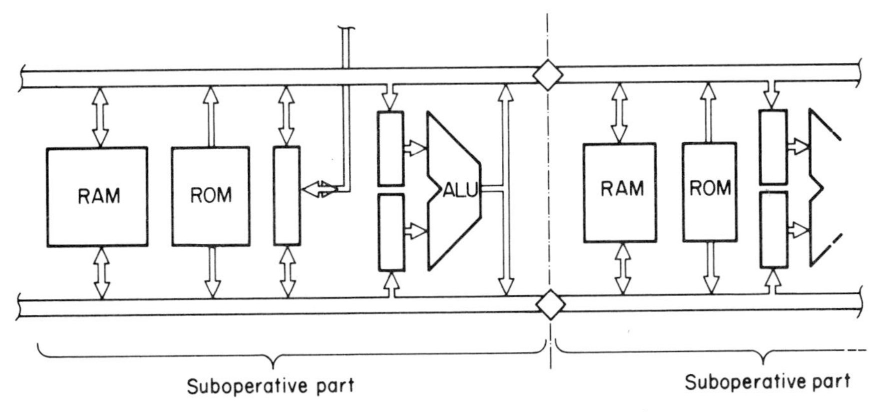

Figure 1 Functional Templet for Data Path.

The functional organization that is suggested for the data path uses a two-bus system, where buses are only cut to defined sub-data paths which may run in parallel (e.g. for data and address computation) (Figure 1).

The electrical behaviour of these buses is the same as the electrical behaviour of a static memory (Figure 2). All sources and sinks of the bus look like static memory points.

In each sub-data path all registers have double access to the two buses. A set of ROM words are used to record the immediate values that are necessary. Input–output connections are linked to input–output registers. Arithmetical and Logical Units received their data from the buses through

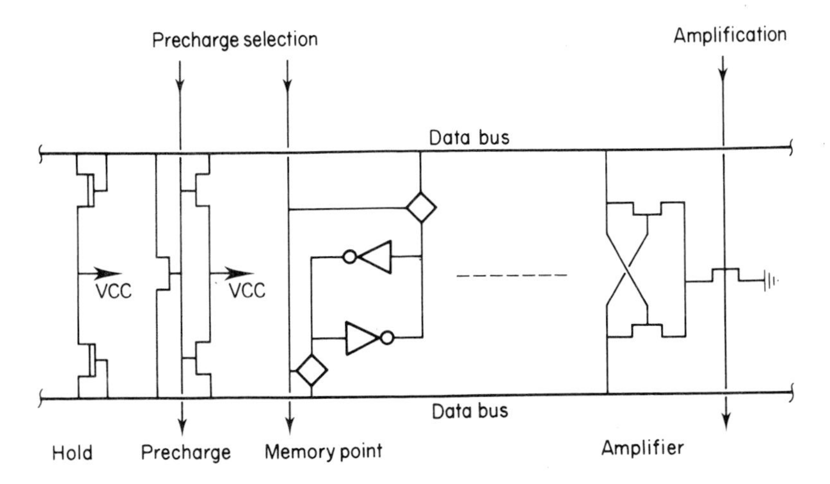

Figure 2 Electrical Organization of a Bus.

operand registers. This functional organization is very powerful. It allows the execution of a register-to-register operation and a register-to-register transfert at every cycle (of two phases). In addition, operands selection and result selection may be used to do multiple-destination transferts.

Example of action instructions executable in one step:

```
PC <= AD <= (RG <= BR) + 00F6,   Q6 <= R12
```

General form of an action instruction set:

```
R1 <=... R2 <= (R3 <=... R4 <= R5 or IV)
               op
               (R6 <=... R7 <= R8 or IV),
R9 <=... R10 <= R11 or IV
```

(IV stands for Immediate Value.)

Such instructions are executed in one step of two phases:

```
<phi1> ALU input 1 <= R3 <=... R4 <= R5 or IV,
       ALU input 2 <= R6 <=... R7 <= R8 or IV;
<phi2> R1 <=... R2 <= ALU output,
       R9 <=... R10 <= R11 or IV;
```

during each phase two transferts are executed using the two buses.

3.2 Data-Path Compilation

The analysis of the set of actions that can be executed by the data path determines how the data-path templet must be adjusted to obtain the real data path of the IC. For this the general form of the executable actions is written as the following grammar:

```
<step> ::= <operation> , <transfert> ;
           <operation> ;
           <transfert> ;

<operation ::= <assigned reg> <operand> op <operand> ;
               <assigned reg> op <operand> ;

<transfert> ::= <assigned reg> <source> ;

<assigned reg> ::= <dest reg> <= <assigned reg> ;
                   <dest reg> <= ;
```

<operand> ::= <source> ;
(<transfert>) ;

<dest reg> ::= reg ;

<source> ::= reg ;
immediate value ;

All action instructions that are recognized by this grammar may be executed in one step by a data path realized according to the templet. The specific features and size of this data path may be obtained during the syntactic analysis of a set of action instructions extracted from an algorithm. A specific data path that is able to execute this set of action instructions must have all the features that are necessary to execute each action instruction. This process is very close to a compilation (Figure 3).

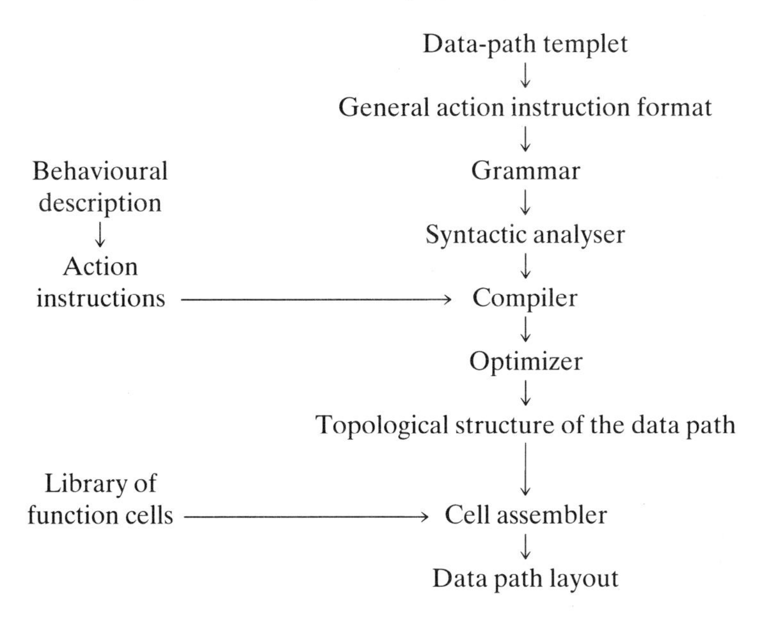

Figure 3 Data Path Compilation.

3.3 Parallelism Into Data Path

The algorithm used to specify the behaviour of the planned IC may invoke several action instructions in parallel at each step. The execution in parallel of these action intructions requires the data path to be divided into several

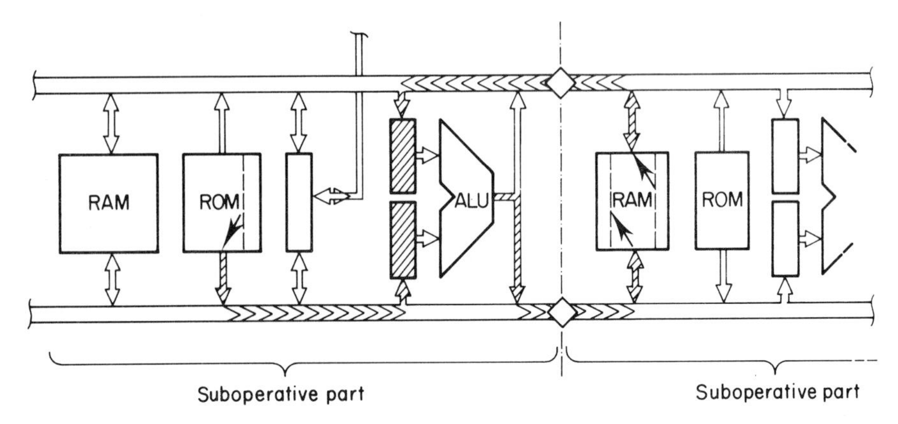

Figure 4 Execution of a Complex Action.

sub-data paths, which may run in parallel. The design of these sub-data paths requires the determination of

(i) the subset of registers to be implemented into each sub-data path;
(ii) the operations that the operator(s) of each sub-data path must execute;
(iii) the possible connections between sub-data paths.

The process for designing sub-data paths starts by splitting the total set of action instructions into subsets of "compatible" action instructions. Each step of the algorithm invokes into each subset a group of action instructions which may be executed by an elementary sub-data path in one step. However, some steps of the algorithm may invoke complex action instructions that need to be executed in order that several sub-data paths be connected together (e.g. an action instruction which fetches operand 1 from sub-data path A, operand 2 from sub-data path B, performs an operation into sub-data path B and store results into sub-data path A) (Figure 4).

In order to perform such action instructions, the sub-data paths must be connected together as a "strip" by bidirectional switches between their buses. The order of the sub-data paths in the data-path strip determines the possible connections that may be used to perform complex action instructions.

3.4 Access to Subfields of Registers

Almost all algorithms that may be written to specify the behaviour of an IC need access to the subfields of registers viewed as bit strings (Figure 5). These features are implemented by duplicating the selection lines of these registers in order to select requested subfields. Extra connection networks

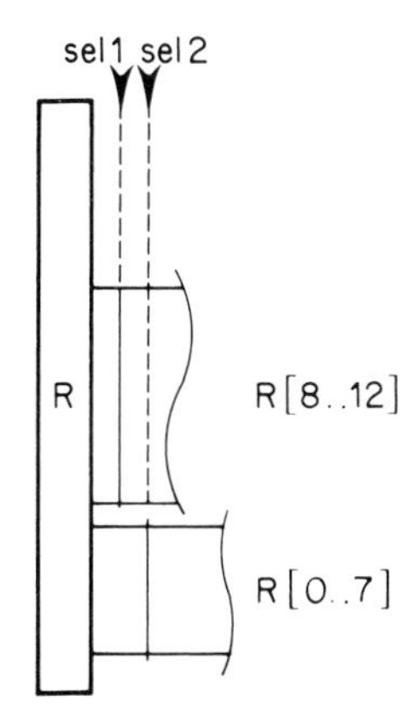

Figure 5 Access to the Subfield of a Register.

must be designed when selected subfields of registers are not connected to the same subfields of buses.

3.5 Data-Path Layout

The layout of the data path generated from this templet is very simple to build. It is obtained by a simple assembly of cells taken from a cell library. The structure of the layout of this kind of data path is fully based on the use of crossed data and control streams. The data stream defines a bit-sliced structure and the control stream defines a function-sliced structure (Figure 6).

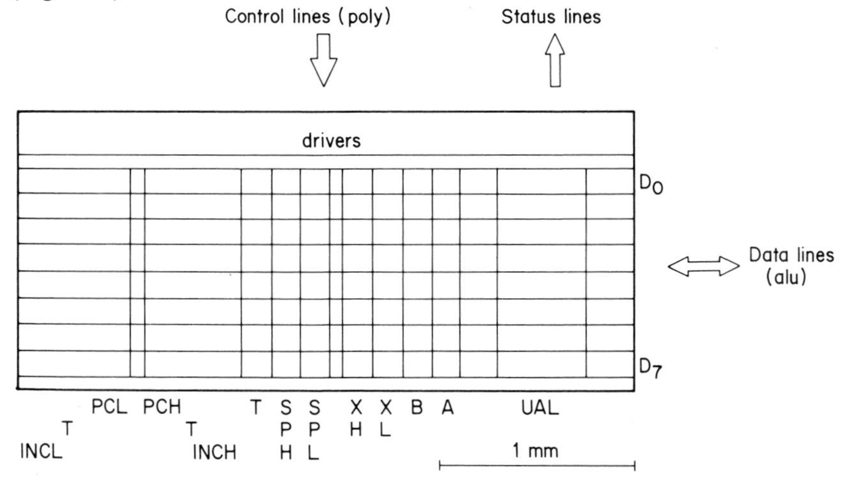

Figure 6 Data-Path Layout Structure.

Because of the use of a non-interrupted double-bus structure in each sub-data path, the function slices may be put in any order to get the layout of a sub-data path. The silicon compiler generates the input of a silicon assembler called LUBRICK [7] as an ordered set of function slices to be assembled and the parameters to generate each function slice. The layout of each function slice is generated by a PASCAL/LUBRICK algorithm, which assembles subcells and takes into account irregularities in the layout (e.g. look-ahead carry propagation, last bit of the bit slice). The connection networks for access to the subfield of registers are also generated by PASCAL/LUBRICK algorithms. These algorithms are very general and are parameterized to generate the layout of the function slices invoked by the silicon compiler. The writing of these algorithms which generate the layouts of function slices is done by experimentally tested IC designers. The use of a standardized templet for data-path compilation is the key that allows the use of a fixed set of generation algorithms to obtain the layout for a full data path.

The LUBRICK system assembles function slices, compacts them into bit direction and routes the connections by extending the "ports" introduced into the layout of each function slice.

4 CONTROL-SECTION DESIGN

4.1 General Considerations

The control structure of the algorithm used to specify the behaviour of the IC constitutes the specifications of the control section. This control structure is given by

(i) the sequencing of the algorithm, which specifies how its steps are conditionally or unconditionally sequenced;

(ii) the conditions that select the action to be performed at each step (conditional execution).

The sequencing of the algorithm is used to design the control section of an IC as an automaton. The conditions are used to modify the command generated by each step of the control automata (parameterization).

The templates for control-section design are just emerging. Several functional templates have already been investigated [8]. But topological guidelines are necessary to get optimized layout.

4.2 Using the Notion of Function Slices

The function slices of the data path may be extended into the control section. Using this technique, the control-section layout is organized as large strips

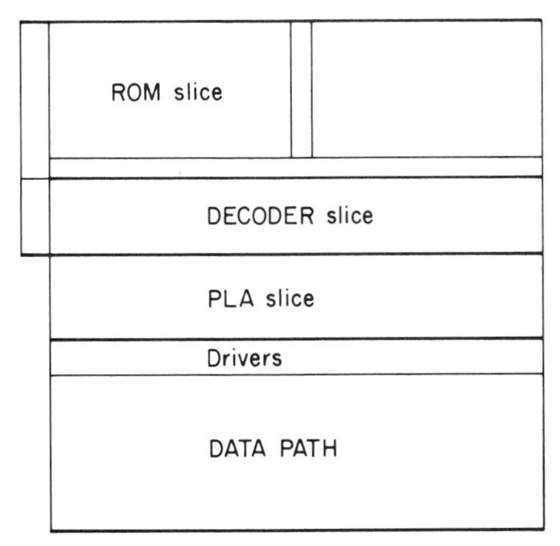

Figure 7 Control-Strip Structure.

parallel to the data-path strip (close to the "strip" of the TEXAS 7000 LSI processor). Each strip of the control section contains two crossed streams (as for the data path). The stream that is parallel to the data path contains timing and condition signals. The stream perpendicular to the data path consists of control lines (it can be divided into function slices). The last strip of the control section (that close to the data path) extends the function slices of the data path. There are three kinds of control strip in the control stream (Figure 7).

(i) *initial* control strips, which generate the control stream. These strips are not upper-connected to other control strips. They receive control and timing signals on sides (e.g. ROM, PLA with classical inputs and lateral outputs).

(ii) *intermediate strips*, which modify a control stream they receive from upper control strips. Two kinds of intermediate strips may be seen:
 (*a*) *validation control strips*, where an initial control stream is validated by timing or condition signals without modifying how it is structured as function strips;
 (*b*) *translating control strips*, where an initial control stream is translated into another control stream (e.g. single matrix PLA, PLA with lateral inputs and outputs);

(iii) *Terminal strips*, which receive a control stream from upper control strips and generate coded information on their side(s). These streams do not output any control stream (e.g. PLA with lateral inputs and classical outputs).

It is often necessary to add some data-path strips to this set of strips. This kind of strip is used to compute addresses for control (e.g. microaddress computation) or parameterization (e.g. computation of register addresses). These strips are often initial or intermediate because they create, or they add, information (microaddress or register address) to the control stream. When they are in the intermediate position they must have a high transparency for the control stream.

4.3 Control-section Compilation

The sequence of control strips depends on the features of the control structure of the algorithm that defines the behaviour of the planned IC.

Control-section templets specify predefined sequences of control strips. They also specify the link between the features of the control structure of the algorithm and each control strip. As an example, each level of interpretation is implemented by a segment of several control strips. The link between two segments depends on whether the commands/instructions that are generated by the upper segment are coded or distributed along the separation between two control strips. The organization of a segment depends on the timing structure of the algorithm. When this timing structure is very simple (few steps in a repetitive order) the control segment is organized as a timing generator and contains

(i) an initial strip that is a decoder receiving conditions (instruction codes and internal conditions) on its side;
(ii) an intermediate strip that is a PLA which validates the initial control stream by timing informations generated by an automaton called a "timing generator" located on the side of the IC.

When the timing structure of the algorithm is complex the control segment may use a control ROM. This kind of segment contains

(i) an initial strip containing a data path to compute the microaddress;
(ii) the strip containing the ROM itself;
(iii) a strip containing microinstruction register and microinstruction fields decoders.

5 CONCLUSIONS

The compilation of an IC that looks like a special purpose microprocessor seems possible in order to get optimized layout. Such a tool could be the key for the VLSI implementation of new architectures of very large ICs containing several millions of transistors. These very large ICs could be built by using microprocessor-like cells as components. The method suggested in this paper for doing automatic design consists of using topological templets which contain the technical know-how accumulated from engineering work. In addition, this approach could be an efficient strategy for pragmatic research.

REFERENCES

[1] D. Johansen, Bristle blocks: a silicon compiler. *16th DAC* (1979).
[2] F. Anceau, CAPRI: A design methodology and a silicon compiler for VLSI circuits specified by algorithm. *3rd Caltech Conf. on VLSI* (March 1983).
[3] R. Gross, Silicon compilers: a critical survey. *Dept of Computer Science, University of North Carolina*, Chapel Hill (May, 1983).
[4] J.M. Siskind, J.R. Southard and K.W. Crouch, Generating custom high performance VLSI design from succinct algorithm description. *Conf. on Advanced Research in VLSI, MIT* (Jan. 1982).
[5] S. Chuquillanqui, T. Perez Segovia, Paola: A tool for topological optimization of large PLAs. *19th DAC, Las Vegas* (1982).
[6] J. Werner, The silicon compiler: Panacea, Wishful thinking or old fact? *VLSI Design* (Sept./Oct. 1982).
[7] J.P. Schoellkopf, LUBRICK: A silicon assembler and its application to data path design for FISC. *VLSI 83, Trondheim* (August 1983).
[8] M. Obrebska, Comparative study of control units of microprocessor using different design methodologies. *Int. Conf. on New Trends in Integrated Circuits, CMU* (April 1981).

5. LSI-processor architecture

F. ANCEAU

1 INTRODUCTION

From several points of view the LSI-processor story is a remake of the minicomputer story, which was in itself a remake of the mainframe computer story (Figure 1). Each of these successive waves has rediscovered, improved, adapted and extended concepts and techniques developed in the preceding waves. The LSI-processor story began in 1972 with the INTEL 4004, and is close now to offering single-chip realizations from the best known mainframe computers. The leading force of this evolution is the continuous improvement of LSI technology. The design techniques and architectural approaches for LSI processor use both techniques and concepts of computer design and of IC design.

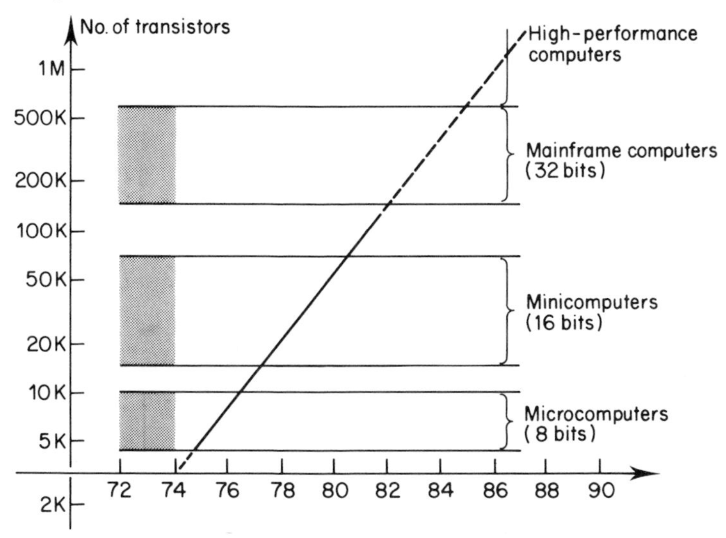

Figure 1 Evolution of IC Technology and Single-Chip Realizations of Computer Ranges.

2 EXTERNAL AND INTERNAL COMPUTER ARCHITECTURE

Since about 1965 the internal architecture of computers has become very different from their visible external architecture. This comes from the differences between external and internal architecture requirements. The external architecture must provide complex and dynamic features and tools for programmers. The internal architecture must provide an optimal trade-off between computing power and cost. Several layers of interpretation are often used to implement the features needed by the external architecture in using the possibilities provided by available technology.

The use of several interpretation levels gives a large independence between external and internal architectures. Exceptions occur when technological constraints are too high (too low complexity or speed available). In these cases, the complexity of the internal architecture is limited. The internal architecture becomes closer to the external architecture. This effect occurs at the beginning of the use of a new technology when starting a new range of computers (vacuum-tube technology which started the mainframe range in the 1950s, IC technology which started the minicomputer range in the end of the 1960s, and LSI technology which started the microprocessor range in the 1970s).

3 COMPUTER RANGES

From a single specification set (external architecture) several internal architectures may be designed. All of them correspond to a local optimization of the cost–performance relationship. When all parameters of the design of the internal architecture have standard values, a "Standard Architecture" is obtained (no features to speed up the computer or to decrease its cost). The Standard Architecture is often the most efficient trade-off in the possible range of internal architectures. In this way it may be used as a reference point. From Standard Architecture low-range computers may be designed from the use of serialization techniques. High-performance computers may be designed by the use of pipelining techniques. However, the economical effectiveness of the low-range and high-performance implementations of computers are less optimal than Standard Architectures. Almost all existing LSI processors use a Standard Internal Architecture. The attempts to design low-range LSI processors have just provided a small decrease in cost but a large decrease in performance. The implementation of high-performance architectures (pipeline architecture) seems more interesting. They have more chance of existing when IC technology will be able to produce them.

4 INFLUENCE OF TECHNOLOGY ON THE EXTERNAL ARCHITECTURE

The influence of the technological features on the external architecture of LSI-processors is as follows.

(i) *Technological limitations*, which act indirectly on the external architecture through the limitations put on the internal architecture.

(*a*) *Complexity limitations*. The relationship between complexities of external and internal architectures is not simple and depends on the complexity of the external architecture, the range of performances requested, and the manner in which chip complexity is measured (silicon area or number of transistors). The increase in design and test costs may lead designers to choose architectural strategies that increase realization costs in order to decrease design and testing costs.

(*b*) *Speed limitations*. The importance of this constraint is now decreasing. Technological improvements will allow feature size and propagation delays to be simultaneously reduced. The experience of the existing LSI processors shows that when the complexity available with technological improvement allows the monolithic realization of a range of computers, the delay time of the internal gates provided by this technology is almost the same as the speed of the SSI and MSI gates previously used to build this range of computers. The most severe limitation of the speed introduced by the technology occurs as a bandwidth limitation on the interfaces between the chips which together constitute a system.

(ii) *Evolution of the application domain*. The continuous decrease in the cost of the LSI realization of computers suggests the use of these machines in new application domains where the use of a computer was unthinkable a decade ago (e.g. car ignition systems, appliances). These new applications introduce new constraints on the external architectures of these LSI computers. In this context it would be interesting to investigate why LSI components for control purposes are LSI realizations of computers instead of having followed a more independent route.

5 INFLUENCE OF TECHNOLOGY ON THE INTERNAL ARCHITECTURE

The technological features that directly influence the main options of the internal architecture are as follows.

(i) the technological features already mentioned as concerning their indirect influence on external architecture;
(ii) the topological contraints linked to the optimization of the layout—the organizations of the main blocks as the crossing area of two streams have a very large influence on the choice of the internal architecture;
(iii) the difference between the rules in measuring the cost of a LSI processor and the rules that were used to measure the cost of a SSI/MSI processor lead to change of design style to preserve optimality.

6 DATA-PATH AND CONTROL SECTIONS

Like any other sequential machine, an LSI processor may be decomposed into two subparts from the logical and layout point of view (Figure 2):
(i) a *data-path section*, which performs local data storage (registers), data transfers (buses), and data transformations (ALU, shifters, ...);
(ii) a *control section*, which animates the operative part in order to perform the requested sequences of operations.

The control section animates the data path by exciting control lines. It takes decisions from the electrical levels on status lines excited by the data path, the input control pins and other internal modules (e.g. interrupts).

The use of several layers of interpretation leads to the dividing of the control section into the stacking of several sub-control sections. Each of them implements a layer of interpretation (Figure 3).

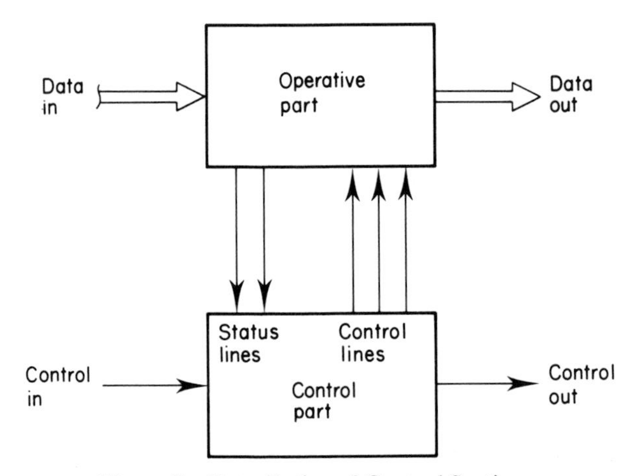

Figure 2 Data-Path and Control Sections.

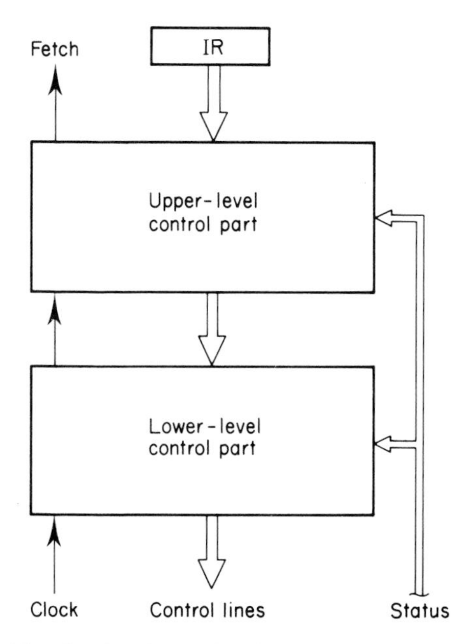

Figure 3 Control Section Implementing Several Layers of Interpretation.

7 EVOLUTION OF THE INTERNAL ARCHITECTURE

7.1 Evolution of the Layout Techniques

The evolution of the LSI-processor architecture is strongly linked to the progressive learning of the efficient use of the LSI technology during the last decade.

The initial step of this evolution has consisted in implementing pre-existing internal architectures. Each module of these architectures was designed as a separate layout block. These blocks are put together and routed to obtain an IC.

The inefficiency of this approach leads the designer to modify the internal architecture to increase the efficiency (e.g. to reduce the routing area). At the same time, the internal organization of the blocks is also changed to make use of the features provided by the IC circuitry (e.g. transmission gates, dynamic memorization).

The last step consists in selecting templates, which provide the best trade-off for designing this kind of IC. The templates presolved also the topological problems that occur during the design of this kind of IC. The use of templates improves the quality, reliability and speed of the design.

The evolution scheme has been followed to completion for the data path, but the design of the control section is midway in this evolutionary process.

7.2 First period (1972–1975)

During the first period of the LSI-processor story, the architects who designed these machines had to try to organize them internally as distributed systems. This idea was very popular in computer architecture at that time. Several minicomputers and mainframe computers were using this organization (e.g. GRI 909, ORDOPROCESSEUR). The main objective of this approach was to decrease the wiring inside processors (wiring between boards for minicomputers and mainframes, wiring between layout blocks for LSI processors). This reduction of wiring came from the use of a single bus to interconnect the mains units of a processor (register file, ALU, instruction decoder, I/O ports, . . .). This single bus introduces a bootleneck effect into the exchange of data inside the processor. This bootleneck effect has been reduced by raising the intelligence of the units in order to reduce the rate where they exchange data.

The distribution of the control among the functional units gives the processor the structure of a distributed system. Several LSI processors were designed according to this methodology (INTEL 4004, 8008, 8080 and 8048 families, ROCKWELL PPS4, EFCIS MOM). (See Figure 4).

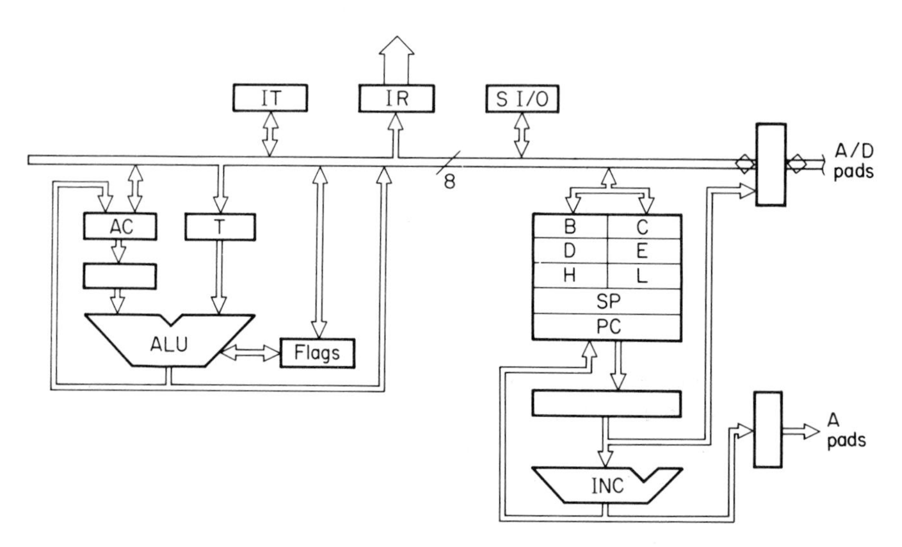

Figure 4 Data Path of the INTEL 8085 LSI Processor.

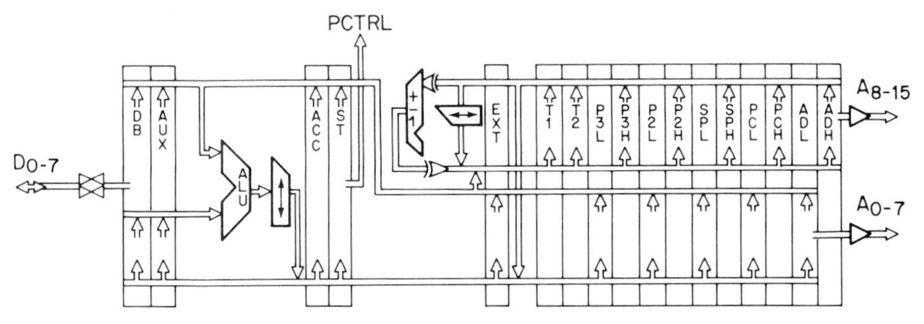

Figure 5 Data Path of the INS 8070 LSI Processor.

7.3 Second Period (1975–1980)

In spite of the good results obtained from this approach, it was discarded. The main reason for this change is that a processor is not a distributed system. Several extra connections have to be added to such a structure to get an acceptable level of performance. The increasing importance of the layout constraints leads designers to find design methods providing better guidelines for making the layout.

The next step in this evolution was the organization of the layout of the data path as bit-slice structures (the first LSI processor using a bit-sliced data path was the Motorola MC 6800). This approach releases the wiring contraint and allows the design of a data path using a large number of internal buses. The use of transistor switches to connect buses gave the possibility of splitting data paths into several sub-data paths running in parallel. The use of a static memory circuitry for register banks (included into the bit slice) gave the possibility of using a large number of registers. (See Figure 5.)

7.4 Final period (since 1980)

The need to reduce the design cost of large microprocessors has led designers to minimize the number of cells used in data-path layout. This standardization of the layout structure involves a standardization of the functional structure of the data path. The first application of this approach was the MC 68000 (Figure 6). The result of this evolution in data-path design may be considered as optimal. It gives way to design data paths which are efficient and easy to design with a high density of layout.

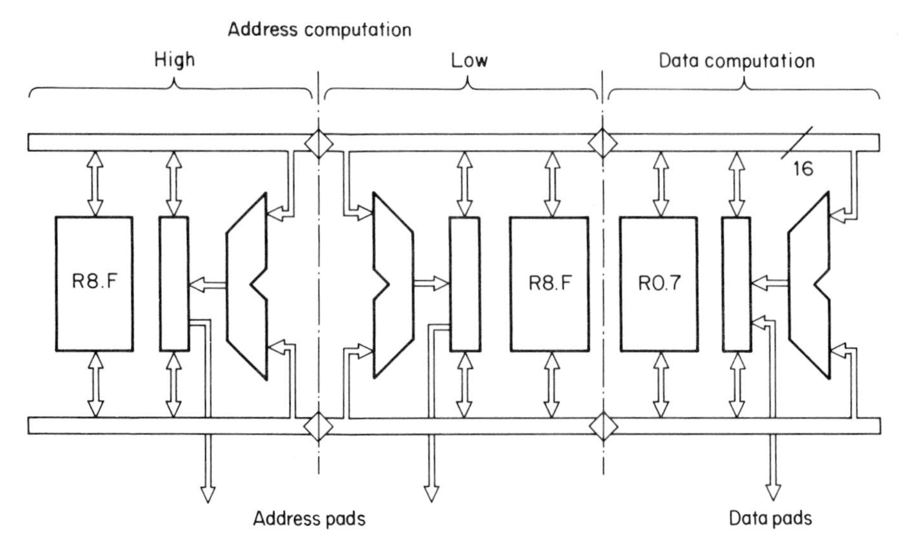

Figure 6 Data Path of the MC 68000 LSI Processor.

7.5 Control-section Design

The architecture of the control section has not yet reached the same level of perfection as that of the data path. A large part of the evolution of the complexity of LSI processors is the increase in the relative size of the control section (except for the RISC-like LSI processors [13]). This means increased complexity of the external architecture of LSI processors is more related to increased complexity of the instruction set (number of instructions, number of the addressing modes) than to the complexity of the operations that may be executed.

Several functional organizations of the control section have been investigated [21]. Some of the main results of this study are

(i) the level of optimization of the control section (in terms of silicon area) increases with the number of conditional actions it uses (MEALEY form of the control automata);

(ii) the command generation and the sequencing mechanisms must be realized by separate hardware units independently adapted to their own function.

Several attempts have been made to design regular layouts for the control sections. One attempt consists in standardizing the floor plan of an LSI processor, which is a good trade-off from the layout point of view (the MC68000 and HP 9000 have similar floor plans) (Figure 7). Another attempt

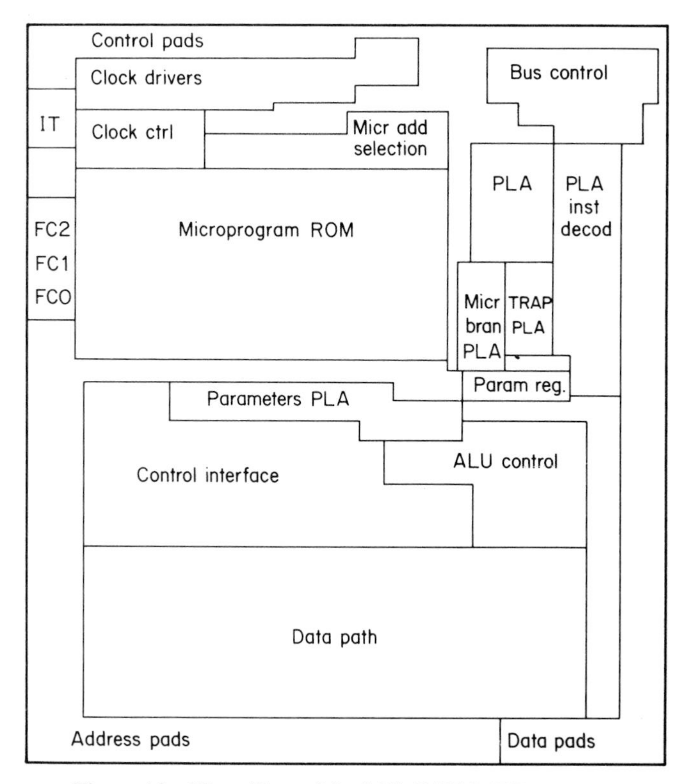

Figure 10 Floor Plan of the MC 68000 LSI Processor.

consists in using a structure of strips parallel to the data path [19] (see also Chapter 4 of this volume, silicon compilation).

8 CONCLUSION

LSI processors are the 'present day' of minicomputers and the 'near future' of mainframe computers. Because their production costs are far lower than those of realizations using previous technologies, their design cost must be not higher (or lower) than the design cost of these previous realizations. Systematic design methods are in development. These systematic methods are also being developed for automatic design in silicon compilation.

The improvements in LSI technology will allow the monolithic realization of all families of existing computers at the beginning of the 1990s. This means that new ranges of computers, directly designed for VLSI realization, must be imagined at this period. The future of computer hardware must be imagined on silicon.

REFERENCES

[1] F. Anceau, M. Marinescu, M. Henry and J.P. Potet, MSQ: a microsequencer as a monolithic component. *EUROMICRO Conf., Nice* (1975).
[2] L.G. Dang, P.B. Askin, R. Yee and M. O'Brien, CMOS/SOS 16 parallel micro CPU. *ISSCC Conf.* (1977).
[3] G. Lowwie, J. Wipfli, E. Ebright, A dual processor serial data controller chip. *ISSCC Conf.* (1977).
[4] J. Beekmans, P. Danielse, C. Fernandes, L. Quere, F. Schiereck and A. Vermeulen, A complete 16 bits microprocessor on a chip. *EUROMICRO Conf.* (1978).
[5] M. Shima, Demystifying microprocessor design. *IEEE Spectrum* (July 1978).
[6] J. McKevitt and J. Byliss, New options from big chips. *IEEE Spectrum* (March 1979).
[7] A. Guyot, M. Henry and M.E. Vergniault, The MOM 400 single chip microcomputer. *ESSIRC Conf.* (1979).
[8] J.W. Beyers, L.J. Dohse, J.P. Fucetola, R.L. Kockis, C.G. Lob, G.L. Taylor and R.E. Zeller, A 32 bits VLSI CPU chip. *ISSCC Conf.* (1981).
[9] W.W. Lattin, J.A. Bayliss, D.L. Budde, S.R. Colley, G.W. Cox, A. Goodman, J.R. Rattner, W.R. Richardson and R. Swanson, A 32 bits VLSI microframe computer system. *ISSCC Conf.* (1981).
[10] L. Kon, A 2 bit microframe computer system. *ISSCC Conf.* (1981).
[11] M. Pomper, W. Beifuss, K.H. Horninger and W. Kaschte, A 32-bit execution unit in an advanced NMOS technology. *IEEE J. Solid-State Circuits* (1982).
[12] J.E. Campbell and J. Tahmoush, Design considerations for a VLSI microprocessor. *IBM J. Res. & Dev.* (1982).
[13] D.A. Patterson and C.H. Sequin, A VLSI RISC. *Computer* (Sept. 1982).
[14] D.E. Blahut, A.K. Goksel, R.H. Krambeck, H.F.S. Law, P.M. Lu, W.F. Miller and H.C. So, The architecture and implementation of a 32 bit microprocessor with minicomputer performance. *Compcon* (Feb. 1982).
[15] K.P. Burkhart, M.A. Forsyth, M.E. Hammer and D.F. Tanksalvala, An 18 MHz 32 bit VLSI microprocessor. *Hewlett-Packard Journal* (August 1983).
[16] F.J. Gross, W.S. Jaffe, D.R. Weiss, VLSI I/O processor for a 32-bit computer system. *Hewlett-Packard Journal* (August 1983).
[17] A.K. Goksel, J.A. Fields, F. Larocca, P.M. Lu, W.W. Troutman and K.N. Wong, The design of a VLSI management chip for the BELLMAC 32 bit microprocessor. *ESSIRC* (Sept. 1983).
[18] M. Pomper, J. Stockinger, J. Augspurger, B. Muller and U. Schwabe, A 300K transistor NMOS peripheral processor. *ESSIRC* (Sept. 1983).
[19] J. Stockinger and S. Wallstab, A regular control unit for microprocessors. *ESSIRC* (Sept. 1983).
[20] F. Anceau, Architecture and design of von Neumann microprocessors. *NATO Summer School, UCL* (1980).
[21] M. Obrebska, Comparative study of control units of microprocessor using

different design methodologies. *Int. Conf. on New Trends in Integrated Circuits, CMU* (April 1981).

[22] J.P. Schoellkopf, datapath design for FISC: a 16 bit P-code microprocessor. *ESSIRC* (Sept. 1983).

[23] C. Piguet, Design methodology for full custom CMOS microcomputers. *INTEGRATION £1* (1983).

[24] R.W. Sherburne, J.R. Manolis, G.H. Katevenis, D.A. Patterson and C.H. Sequin, Datapath Design for RISC. *Conf. on Advanced Research in VLSI, MIT* (Jan. 1982).

Index